Advance Praise

Fr. Michael Chaberek is a must-read author when it comes to science and faith. He consistently keeps in mind the proper roles and limitations of science, philosophy, and theology. Fr. Chaberek is deeply steeped in the Catholic intellectual tradition—from the Fathers, to St. Thomas Aquinas, to the twentieth-century greats—while at the same time being more sensitive to the advances of modern science than almost anyone in the dialogue about science and religion. I cannot recommend his work highly enough.
—Logan P. Gage, PhD, Franciscan University of Steubenville

Thanks to scholars such as Fr. Michael Chaberek, OP, a new way forward is opening up in the old, sterile debate between creationists and evolutionists. In this volume, Fr. Chaberek enables readers who are not specialists in the field to grasp the essential features of this new approach, combining insights from: Progressive Creation theory; Intelligent Design theory; responsible Scripture scholarship; and faithful adherence to the Catholic Tradition. By drawing fresh insights from these sources—as well as recovering some vital ones too long forgotten—Fr. Michael breaks through the impasse between older, entrenched positions, and reaches instead for a new vision of how God created and fashioned our universe, and life on planet earth. Fr. Chaberek has launched us on our way in this book. Those who read it with an open mind will find themselves enlightened, challenged—and grateful for his guidance.
—Robert Stackpole, STD, Director Emeritus,
John Paul II Institute of Divine Mercy

As soon as Pius XII permitted Catholics learned in both theology and human sciences to consider the possibility that the human body might have arisen from pre-existent and living matter, the learned threw all caution to the wind, ignored the Pope's caveats, and embraced theistic evolution. Never mind the overreach of Darwinism beyond microevolution. Never mind the jettisoning of nearly two thousand years of careful commentary on the Bible and theological reflection. Today, theistic evolution is still dominant among Catholic thought leaders. But, as a science, Darwinism is a failure. And its acceptance makes nonsense out of some fundamental Catholic teachings: original sin, God's creation of man and woman, and the very notion of human species—and more.

Fr. Chaberek points out all of these problems in his very readable *Creation or Evolution? A Catholic Dilemma*. It is a testament to the weakness of theistic evolution that its proponents make it very difficult to get these clear criticisms published. Catholics who instinctively know that there is something fundamentally flawed with theistic evolution nevertheless have to work very hard to find the details for themselves. So they should be grateful to Fr. Chaberek for clearly setting out the problems and giving them confidence that the perennial teaching about the creation and fall of Adam and Eve is true.

Fr. Martin Hilbert, PhD, author of *A Catholic Case for Intelligent Design*

As an unabashed and "politically-incorrect" Catholic scholar, Fr. Chaberek presents solid Thomistic philosophical and theological arguments to support the *philosophy of being*. But his fight is a *Cervantean* one, that of a philosophical Don Quixote courageously tilting at the giant windmills of post-Darwinian thought which for 150 years have incessantly harnessed the pervasive metaphysical flux of a Heraclitan *philosophy of becoming*. The result is a lively and lucid analysis of the metaphysics of creation, evolution and theistic evolution, as well as an introduction to the scientific theory of Intelligent Design (ID).

Does Man have eyes *for seeing* (creation) or does he see *because he got eyes* (evolution)? For anyone who would like to ponder the metaphysical answer to that question, I highly recommend this authoritative and luminous book.

—Marc Mullie, MD, Ophthalmologist

Fr. Michael Chaberek is offering fascinating information and suggestions that are very hard to find elsewhere. His research provides us with original insights very much needed in the Church right now—and in society as a whole.

—Bruce Chapman, co-founder of Discovery Institute, former U. S. Ambassador to the United Nations

Creation or Evolution?
A Catholic Dilemma

Creation or Evolution?
A Catholic Dilemma

Fr. Michael Chaberek, O.P.
In conversation with Steve Greene

STENO
INSTITUTE FOR FAITH & SCIENCE

in partnership with

Inkwell
PRESS

Description: To reconcile faith and science, is it enough to say "God could have used evolution"? Many Christians adopt this theistic-evolution point of view. It is now treated as the quasi-official Catholic position. Yet can it really harmonize Scripture, theology, and modern science? How does it align with the Bible read through the Church's long tradition and in light of classical Christian philosophy? Does it resolve the tension between Genesis and the Church Fathers on one side and the grand evolutionary narrative embraced by atheists on the other? And what role should intelligent design play in Catholic teaching? *Creation or Evolution? A Catholic Dilemma* presents a searching dialogue between two Catholics—a layman with long pastoral experience and a Dominican priest-scholar specializing in the faith-and-science debate—who probe whether theistic evolution remains a viable option for believers or whether creation, rightly understood, offers the sounder foundation for understanding life, humanity, and the cosmos.

Imprimatur: Łukasz Wiśniewski, OP, Prior Provincial of the Polish Dominican Province, hereby grant the Imprimatur to *Creation or Evolution? A Catholic Dilemma* by Michael Chaberek, OP, Reg. Prow. 721/25.

Library Cataloging Data
Creation or Evolution? A Catholic Dilemma by Fr. Michael Chaberek
Library of Congress Control Number: 2026930114
ISBN: 979-8-89946-017-3 (Paperback), 979-8-89946-018-0 (Hardcover), and 979-8-89946-016-6 (eBook)
BISAC: REL122000 RELIGION / Creation
BISAC: PHI013000 PHILOSOPHY / Metaphysics
BISAC: SCI075000 SCIENCE / Philosophy & Social Aspects

Book Cover Design: The cover image was created by Inkwell staff using artificial intelligence.

2321 Sir Barton Way
Suite 140-1032
Lexington, KY 40509
press.inkwell.net

c/o Ordo Iuris International Foundation
1120 Avenue of the Americas FL4
New York, NY 10036, USA
stenoifs.org

The first formation of the human body could not be by the instrumentality of any created power, but was immediately from God.

Saint Thomas Aquinas

Table of Contents

Introduction

In 1859, Charles Darwin's *On the Origin of Species*[1] sent shockwaves through all the intellectual institutions of the Western world, including the Catholic Church.

Although there was much debate in the Church in Darwin's time over what the implications of his evolutionary hypothesis might be vis-à-vis Christian doctrine like the existence of a Creator, the inerrancy of Scripture, and the creation of human beings in God's image, by the mid-twentieth century most Catholic intellectuals had decided that—as long as God could be inserted somewhere in the narrative—evolution was more or less compatible with the Catholic Faith.

But what if there is more to the story?

As scientific discoveries in multiple disciplines have called the viability of evolutionary theory into question, a growing number of Catholic scholars have re-opened the question of whether Catholicism and evolution really are so compatible, after all. One of the leading lights of this new movement is the author of this book, Fr. Michael Chaberek, OP.

Fr. Chaberek is a proponent of a new faith–science synthesis known as Progressive Creation. Progressive Creation (PC for short) seeks an alternative to the false dichotomy of Darwinian evolution or fundamentalist creationism, so often presented as the only options for those grappling with evolution. PC synthesizes the latest science with the best of the Catholic philosophical and theological tradition to offer a view of the origin of life, biodiversity, and ultimately man, which is fully compatible with the teaching of the Catholic Church.

In the meantime, the evolutionary framework for addressing the origins questions is in serious trouble scientifically.

Whether it be the Cambrian explosion, the discovery of DNA's information-bearing sequencing, the organelles and molecular machines in every living cell, or the irreducible complexity of countless features of living beings, the Modern or Extended Darwinian Synthesis, popularly known as neo-Darwinism, is everywhere failing in spectacular fashion to explain what the hard sciences have now observed.

Far from serving to confirm the neo-Darwinian hypothesis that natural selection, acting on random and genetic variations, can explain the rise of life and biodiversity, advances in nearly every branch of the empirical sciences, and the staggering expansion of what we have learned about life in all its variety has, instead, revealed Darwinian evolution to be a failed hypothesis.

It was inevitable that in a post-Christian culture such as ours, some fundamentally secular paradigm, like Darwinism, would arise to provide the new framework for understanding reality. It is a necessity for a materialist worldview. As atheist biologist Richard Dawkins said in his book *The Blind Watchmaker,* "Darwin made it possible to be an intellectually fulfilled atheist." If one wonders why modern man treats Darwinian evolution like a secular religion, one need look no further than that statement. It gives the game away. Darwin provides the essential element previously missing for the post-Christian secularist: a purely natural explanation of biological reality, a plausible mechanism for how we might have all got here, with no meaningful reference to God.

Darwin's hypothesis shook both science and society because, if true, it meant that nature alone could account for everything in nature; that the natural world was self-explanatory and self-sufficient. This is precisely why Dawkins embraces Darwin: Darwin quietly did away with God.

This plain fact makes all the more baffling the widespread tendency of Catholic academics and intellectuals in our time to not only embrace Darwinism, but to run to its defense—albeit while insisting on the obligatory caveat of God's existence and indirect involvement—even in the face of its manifest and multiplying scientific shortcomings, and its fairly obvious points of departure from the Catholic philosophical and theological tradition.

This attempt on the part of Christians to accommodate Darwinism goes by the name of Theistic Evolution (TE)—an umbrella term broadly indicating a school of thought which holds that God exists and created the universe at its origin, but then left evolution, which He influences only indirectly, to naturally produce all life forms. This is now the reigning view across almost all Catholic intellectual institutions.

However, it was not always so. In his 1950 encyclical, *Humanae Generis* [On the Human Race], Pope Pius XII gave Catholic scholars a carefully circumscribed permission to discuss the possible evolution of the human body. The clear limits to this permission, it should be noted, have been flagrantly transgressed by Catholic scholars in the ensuing decades. In any case, most in the TE camp base the validation of their argument that Catholicism and evolution are compatible by referring to this document. Fr. Chaberek, on the other hand, counters by offering us the invaluable service of re-introducing into the conversation the clear and unanimous Church teaching, from the Church Fathers to Darwin's time, which emphatically establishes the orthodoxy of both a literal, historical interpretation of the Genesis creation account, and the special creation—direct, supernatural, divine—of man, of both his soul *and* his body, by God.

So, the question is: if the orthodox Catholic teaching has always been fundamentally incompatible with Darwinism, how is it that Theistic Evolution has so entrenched itself in Catholic academia as to be virtually unquestionable? The idea of God's using evolutionary processes—in which there is no clear or coherent account of how God is present, active, or involved—is deeply problematic philosophically, theologically, and scientifically. And yet the overwhelming majority of Catholic intellectuals seem to view it as all but certain. Why? What is its appeal to Catholics? The answer must ultimately be given by each individual, of course, but there are at least two discernable factors that may shed some light on those questions, in general.

The first is the Church's Galileo hangover. Four hundred years on from the Galileo affair, it still stands in the popular imagination as Exhibit A of the alleged war of religion against science. In consequence, too many Catholic scientists and intellectuals are haunted by the fear that, should they oppose the predominant evolutionary paradigm, they too will find themselves "on the wrong side of history." The thought that they could be cast as benighted, anti-science villains, like those who silenced Galileo, is simply too much to risk. Far better to play nice and blend in.

The second factor is an analogy to the emperor effect. In ancient Rome you were required to offer a pinch of incense on the altar of the emperor, thereby demonstrating your loyalty and tacitly acknowledging his claim to divinity and right to rule. Make the offering and you were free to go about your business. Refuse and you became a dissident, forfeiting your freedom to participate in the life of the empire, and risking prison, exile, or even death.

This is analogous to the situation academics now face in the modern West. Science is the new Roman empire, universal and all-encompassing, claiming all the civilized as its citizens, and Charles Darwin is its god-king. If you wish access, status, freedom, citizenship in this intellectual empire, and all the rights pertaining thereunto, you must simply bend the knee and pledge your loyalty to the emperor. Refuse, and you will find yourself ostracized, exiled, persecuted—and maybe even executed—professionally and academically.

Simply stated, today "I believe in evolution," is the password that opens the doors of academia and the life of the professional intellectual. If you let slip that you have doubts, or are found to be among the unbelieving, you are likely to find yourself an outcast, whose invitation to the party seems always conveniently to be forgotten by those hosting it.

Nevertheless, given the desperate state of neo-Darwinism on its dwindling scientific merits, and the conspicuous absence of any viable materialist alternative, it appears as though the Catholic intelligentsia have chosen to lash themselves to the mast of a sinking ship. Ironically, it seems likely they will end up on the wrong side of history, in spite of their herculean efforts to avoid precisely that fate.

Our conversation with Fr. Chaberek is a welcome alternative to this perplexing hybrid of recalibrated faith with a waning scientific hypothesis. *Creation or Evolution?* introduces us to the critiques of Darwinism and Theistic Evolution coming from a growing number of philosophers, theologians, and scientists. This book illustrates beautifully how those critiques respond to the myriad questions that surround the synthesis of Christian faith and biological macroevolution.

In these pages, Fr. Chaberek also offers a clear explanation of the competing views of evolution, and an extremely helpful explanation of the scientific theory of Intelligent Design (ID) and how it reinforces Catholic orthodoxy. On all these topics, he gives the reader clarity and insight, speaking always from the heart and mind of the Church.

When all is said and done, *Creation or Evolution?* is a much-needed invitation to a dialogue that should have begun in earnest long before now, especially among Catholics and Church institutions, but which heretofore has all too often been the sound of one evolution-promoting hand clapping.

Finally, this book is not just for Catholics. It is also intended for anyone genuinely interested in the question of origins, and for anyone sincerely seeking the truth. For those dogmatically committed to the evolutionary paradigm and the dismissal of its critics, it will likely be nothing more than an irritation, but for those open to what a rational integration of the latest science with the best of the Western philosophical and theological tradition might have to offer, it may well be a revelation.

If this book contributes to launching an open dialogue among those interested in these questions, it will have served its purpose. If it simply proves thought-provoking to readers who had not previously considered this issue in all its complexity, that will be success, as well.

May this be for you the beginning of a joyful and inspiring adventure of both faith and reason.

Steve Greene
Director, Holy Family Institute of Catholic Faith and Life
Host of the *Reasonably Rational* YouTube podcast

1. The Evolution Controversy

We are going to talk about evolution in the context of Christian faith. To begin, I want to ask you what, exactly, do we mean by "evolution"? And what is the evolution controversy about?

It is a very important question to begin with. Evolution is quite a concept and different people understand it differently. If all parties adhered to one definition, we would surely avoid many disagreements. First, we should notice a very general division: The notion of evolution may be applied to nature or culture. These two realms are ruled by somewhat different laws so that evolution means something different when considered in these two realms. The controversy which we are going to address belongs to evolution in nature. Therefore, we will not talk about the evolution of political doctrines, economic systems, civil laws, human language, or anything that is a product of the rational activity of humans.

Our subject concerns evolution that occurs in the natural world. But even here we need to narrow down the scope. In the natural world one may think of, for instance, cosmic evolution, that is, how the planetary systems came about, or how the stars burn out over the course of billions of years. One may also consider chemical and biochemical evolution, that is, how organic compounds form from the elements and then how life emerges from those compounds. Our main focus is on biological evolution, that is, the type of evolution pertaining to the development of living organisms.

In biology there is a distinction between micro and macroevolution. How do they differ? Are both types of biological evolution controversial?

Microevolution refers to the variation and modification within species, such as environmental adaptations, bacterial resistance to antibiotics, or changes in the proportions of individuals with certain traits in populations. A standard textbook example of microevolution is the change in the proportion of dark-colored peppered moths to light-colored peppered moths as a response to soot occurring on tree trunks in industrialized regions (so-called "industrial melanism"). Another example is the adaptation of foxes, bears, and other animals to polar, desert, or mountain climates. Another example might be the domestication of an animal.

Microevolution may entail quite dramatic alterations of external characteristics, such as the color or thickness of fur in mammals, or of less-dramatic but still-visible traits, such as the size of beaks in birds, or of internal characteristics that may be invisible externally. For instance, some people acquire resistance to malaria due to a single mutation in the gene coding for the production of hemoglobin in their blood. Another mutation allows some people to digest milk while others cannot. In such cases some intrinsic trait of an organism is strengthened or weakened by genetic mutations, but the external appearance and the form of the organism remain entirely unchanged.

In contrast, macroevolution means that alterations go so far as to change one kind of animal into another kind of animal, as is supposed to happen in the transformation of reptiles into birds or mammals, or of fish into amphibians. In the broadest sense, macroevolution (so-called "monophyletic evolution") is the idea that all living species have a single common ancestor. In this sense, macroevolution is a general statement about the whole animated world, which says that if we went back in time, we would ultimately get to a single common natural ancestor of all known forms of life.

In the debate that we take on here, microevolution does not cause much controversy—both Darwinists and their adversaries agree that microevolution occurs. We can observe microevolution in natural conditions as well as in artificial breeding. In fact, artificial breeding consists mainly of humans' intelligent manipulation of microevolutionary mechanisms. The bone of contention is the claim that these small microevolutionary adaptations, if accumulated over time, would lead to such fundamental changes as, for example, between a lizard and an eagle or an ape and a human. Therefore, the kind of evolution which is the object of our attention here may be called "biological *macro*evolution."

You seem to take the division between micro- and macroevolution for granted. But there are some who dismiss that division as unscientific. Why do they make that claim?

The very division was invented by evolutionists themselves. Today, however, some scholars reject the distinction between micro- and macroevolution because, they think, we cannot specify the limits of microevolution. It is almost like saying that one cannot distinguish between red and green, because the color palette contains intermediate colors, such as orange and yellow, that transition smoothly from one into another. Yet, those biologists who deny the distinction between micro- and macroevolution admit that there is a difference between a lizard and a bat, and that this difference is of a different character than a difference between, say, a wolf and a dog. Therefore, they also understand what micro- and macroevolution are. They are often reluctant to acknowledge the distinction, because they know that evidence for microevolution does not make up for the gap in evidence for macroevolution. Regrettably, instead of finding evidence for macroevolution, they choose to deny the very distinction and thus render the debate impossible. It is as if they tried to nullify the results of a contest because their favorite did not win.

So, can the limits of microevolution be precisely defined?

First of all, we need to ask how precise a definition we need to make our debate possible. The examples of differences between a wolf and a dog on the one hand and a lizard and a bat on the other do not precisely delineate micro- and macroevolution, yet they are sufficient to pursue the debate. Note that even in the strict sciences we sometimes employ imprecise terms simply because such is the nature of our cognition. We need to resort to different levels of abstraction in order to express different ideas and concepts. Surely, the distinction between micro- and macroevolution may be defined more precisely than in the given example. But my point here is that we should not be misled by the false logic of the argument saying that because the distinction between micro- and macroevolution is not quite exact our debate over the limits of microevolution is impossible or futile.

Is there any general criterion to distinguish between micro- and macroevolution?

We may say that the microevolutionary changes are those that differentiate organisms within one genus or—at the furthest—one family. For instance, one horse family (*Equidae*) includes zebras, donkeys and horses. One dog family (*Canidae*) includes wolves, foxes, jackals, coyotes and domestic dogs. So, microevolution may lead to differences such as the one between wolves and domestic dogs. And these changes, as I already said, do not create a controversy. Even the most ardent opponents of macroevolutionary theory allow changes at the level of genus or family. The idea of extreme "species-fixism" is virtually nonexistent in the current debate. Remember that no example of a biological change restricted to one family is an example of macroevolution; therefore, such an example cannot be the evidence for macroevolution. The opponents of the distinction between micro- and macroevolution usually either point out that there are different biological classifications (and we do not know which one is authoritative), or else present a difficulty in classifying a particular organism to a proper taxonomical group. But scholars have always classified organisms. Classification is not just an illusion or an entirely arbitrary activity for hobbyists. There is something objective in nature, in biological entities, that allows us to confidently distinguish fish from amphibians, reptiles from birds, or mammals from plants. In fact, the very skeptics of the distinction between micro- and macroevolution turn on their skepticism only when it comes to evolution. Once the argument over evolution ends, they usually do not have problems with recognizing taxonomic categories. Skeptics simply think that one wins the game by questioning its rules. I don't think that is a fair approach to the debate.

Most people understand the controversy about evolution as a contro-versy concerning the origin of species. How can we understand this debate in the light of the distinction between micro- and macroevo-lution?

The theory proposed by Darwin was aimed at explaining the origin of all species, that is, the entire biodiversity that has ever existed and currently exists on earth. In the context of the division into micro- and macroevolu-tion, we see that the debate does not concern the origin of species as they are understood in contemporary biology. Rather, it is about the origin of broader categories, such as genus and family. These broader categories may be also called natural species. An example of natural species includes dogs, cats, horses, elephants, eagles, butterflies, spiders, etc. The essence of Darwin's theory is that all species, taken as natural species, descended from one or a few ancestors which are the natural progenitors of all living creatures. And this is the disputed claim.

Some think, though, that the term "species" is unclear and flexible.

Such was the thinking of Darwin himself. Had he granted that species were definable and permanent, he could not have proposed that one species transforms into another. The main book by Darwin was entitled *On the Origin of Species.* But what would be the sense of writing a book on the origin of species if there were nothing like "a species" in nature? After all, if species did not exist, the claim that Darwin explained their origin could not be true. This is why Darwin's idea of species is inconsistent—on the one hand he denies their real existence and permanence in order to introduce the idea of transformation and, on the other, he postulates their existence and stability in order to claim that he has explained their origin.

What does Darwin's theory actually claim?

Darwin thought that the mechanism leading to biological macroevolution consists of random variation and natural selection. His theory may be summarized in five points: 1) Each organism generates more offspring than can survive. 2) The newborn organisms differ from each other. 3) Excessive fertility leads to competition and the struggle for survival. 4) The struggle for life results in the *survival of the fittest,* which means that the organisms that randomly acquired some competitive advantage would more likely survive and reproduce. 5) The accumulation of the beneficial traits in subsequent generations, over long periods of time, will result in the diversification of life into all the forms we know. Mind that the first four points are hardly controversial because they—more or less accurately—describe the mechanism of microevolution. The bone of contention is just the fifth point, that is, the claim that if Darwin's mechanism works long enough it "discovers" all biological devices such as the eye, the wings, the brain, or thousands of others at the molecular level.

What is neo-Darwinism?

It is a modification of the Darwinian mechanism of evolution that was introduced after the discovery of genes and the laws of heredity. Darwin did not know where the new traits necessary for natural selection to work came from. He also didn't know about population dynamics that could lead to the fixation of traits regardless of their fitness (genetic drift). Neo-Darwinism postulates that variation (differences in traits in a population) is the result of random genetic mutations passed on to posterity. You see that Darwin's theory, including the neo-Darwinian variant, works through the interaction of chance and necessity—the chance factor being genetic mutation and the necessary factor being natural selection.

Because the neo-Darwinian mechanism of evolution is the one commonly accepted in biology, terms such as "biological macroevolution," "Darwin's theory," or "Darwinism" are usually used interchangeably with the term "neo-Darwinism." However, in our discussion of biological macroevolution the mechanism of evolution is not so relevant. Instead, we focus on the supposed *effects* of this mechanism, that is, the production of new genes, new functional organs, new body plans. and, ultimately, entirely new forms of life. We may therefore conclude that the controversy about Darwinism concerns the question of whether or not random variation, genetic drift, natural selection, and other random and necessary factors can explain the origin of all species. The controversy over biological macroevolution concerns the question of whether the idea of all species descending from a single ancestor, the so-called Darwinian tree of life, depicts the true history of life or, perhaps, species did not evolve from one another but began independently.

Is there progress in evolution?

A majority of contemporary Darwinists insist that there is no progress in evolution. This process has no purpose, no end, and it does not lead to any perfection whatsoever. In their view, evolution consists of merely differential reproduction and survival of individuals with different genetic traits. But these same evolutionists firmly claim that all organisms—both currently alive and extinct—share common ancestry. So, the effect of the process that lacks "any progress or goal" is the emergence of multicellular organisms from the single-celled, eukaryotes from prokaryotes, amphibians from fish, reptiles from amphibians, birds from reptiles, mammals from reptiles, and humans from apes. Thus, the same process that has neither goal nor progress infallibly leads from simplicity to complexity in biology. I bet you see how this reasoning lacks coherence.

Surely, we may define progress in many different ways. But let's stick to a definition which is strict and empirically verifiable; let progress mean an increase in complexity in biology, that is, a growth in the number of distinct functions and organs. It is obvious that between bacteria and mammals there is a huge leap in complexity. Therefore, either evolution did not go from bacteria to mammals or there is progress in evolution. Evolutionists claim that evolution *did* go from bacteria to mammals; therefore, they cannot consistently maintain that there is not any progress in evolution.

Why do evolutionists insist that there is just change but no progress or goal in evolution?

The reason they do not admit progress or direction in evolution is that had they admitted it, they would have needed to explain the cause of it. One of the basic principles of being and reasoning says that everything that exists has a sufficient reason for its existence. In particular, this means that a lower cause cannot bring about a higher effect. For example, a wagon which does not have an engine does not pull a locomotive, but a locomotive with a functioning engine pulls a wagon. Therefore, a locomotive that possesses a perfection in the form of a working engine can be a cause of the movement of the wagon which does not possess that perfection. This does not work the other way around. The wagon simply does not have that "organ" and the function that is performed by the engine in the locomotive. This is why it cannot pull. In the evolutionary view of nature we encounter a similar problem. We do not know how, for instance, a plant would change into an animal since the latter has more organs and functions and therefore is more perfect. Introducing many intermediate stages between plant and animal does not help the case. In fact, it complicates the transition, because arriving at each intermediate stage would require another infallible step "towards an animal." But the more steps there are to make, the more difficult it is to proceed blindly in one direction. Therefore, saying that there are many intermediate stages in evolution (gradualism) actually reinforces the claim that there is direction and progress in evolution.

Does the principle of sufficient cause force us to acknowledge progress in evolution?

According to the principle of sufficient cause, there could be a contrary process: For example, an animal could transform into a plant. But in that case we would not speak about evolution but "devolution," that is, a reduction of organs and functions. The phenomenon of devolution is actually observed in nature, for instance, when organs lose their original function and become so-called rudimentary or vestigial. Oftentimes, evolutionists present these organs as evidence for evolution. In fact, however, rudimentary organs testify against the creative power of evolution. They confirm that there occurs an opposite process to evolution which is devolution and reduction of biological diversity which in the beginning must have existed in some perfected form. The problem with the general evolutionary vision of life is this permanent need of a higher effect being generated by a lower cause. Evolutionists realize that Darwin's mechanism does not explain why we see progress in the history of life. But, instead of acknowledging the insufficiency of the Darwinian mechanism, they choose to deny even the most obvious fact that if the "tree of life" is true, there must be progress and perfection in evolution.

Some Christian scholars would admit that Darwinian evolution is blind, but because evolution is guided by God it can achieve its ends. A popular term coined by these Christian proponents of biological macroevolution is "theistic evolution." There are a number of Catholic scholars, as well, who support this concept. What is theistic evolution?

According to this view, God used evolution in forming the natural world, specifically in creating different species of living beings. In theistic evolution, evolution is called the secondary (technically, the instrumental) cause of creation. As, say, a sculptor uses a chisel to form wood or stone, similarly God employed natural processes, such as genetic mutations and natural selection, to produce all species of living beings. Therefore, theistic evolution is an attempt to incorporate evolutionism into Christian doctrine on creation.

Is theistic evolution a viable hypothesis?

There are many problems with this view that I hope to explain in our further conversations. Here I will just point out that theistic evolution differs substantially from the traditional Christian understanding of creation. Theistic evolution does not allow direct divine causality in the formation of the universe. Thus, in theistic evolution the idea of direct creation is replaced with the idea of God working through secondary causes alone. Yet, direct divine causality was the novelty that Christianity brought into the pagan mythological worldview at the beginning of the Christian era. One can say that in theistic evolution we see a revival of the old pagan concepts of emanationism and pantheism, in which it is not God Himself who creates the world, but the lower beings which are empowered by God to create.

2. The Genesis Creation Account

Atheists typically say that one cannot believe in the Bible and appreciate science at the same time. For example, Richard Dawkins claims that a literal reading of the creation story is "inconsistent with our knowledge of the universe's age, or of how living organisms are related to each other." According to Dawkins, Genesis "has no more special status than the belief of a particular West African tribe that the world was created from the excrement of ants." Most atheists view the Bible as no more than a literary fiction of a primitive ancient tribe. What would you say to those who hold this opinion of the Bible?

What strikes many converted scientists is that when they read the Bible for the first time, they do not really find in it anything contrary to the science of their time. Astrophysicist Hugh Ross, for one, when he encountered the Genesis account of creation, was surprised how well it fits within modern cosmology. I think there is no need to look for "science" in Genesis. But if there was a contradiction, believers would be in real trouble. What is needed is the mere lack of contradiction between the biblical and the scientific data and facts. This is enough to remain a serious believer who accepts science. Surely, many atheists, but also theistic evolutionists, would point out a number of difficulties. However, I have no doubt that most, if not all of them, have been already resolved. I do not see anything in Genesis that would contradict modern discoveries. And I mean "facts" and hard "data," not the interpretations and theories built upon them.

This is what distinguishes Genesis from other ancient cosmologies—they all failed the test of modern science by being flatly incompatible even with our basic knowledge about the universe. Genesis alone stands out as the only account that does not contain anything contrary to the facts of nature.

What is the place of the the Holy Scripture in the pursuit of truth?

The Holy Scripture is not a regular book or just one of the monumental works of ancient literature. It is not comparable to the writings of secular sages, whether those from the ancient past or our times. The Bible is the book of life that was created under divine inspiration. Hence, it is the word of God himself addressed to people. What I am saying should be obvious to any Christian. However, if we look at how the Bible is looked upon in today's academic circles, it is hard to believe that scholars recognize its special status, no matter whether they are believers or not. We can say that there is just one principle of biblical hermeneutics—I would call it the "principle of the mercy of the word." It means that God gave us biblical revelation in order to tease us out from ignorance and darkness, and somehow enrich us with truth and wisdom. After original sin we lost something; God became remote and unknown. But thanks to His revelation we have access to the mystery of His existence, internal life, and His involvement in the universe. Thanks to His words contained in the Bible, we transition from the darkness of ignorance to the light of knowledge and understanding. We are transformed in order to participate in the happiness of God. This is how Divine mercy is realized through Scripture. God is merciful and this is the ultimate reason for His self-revelation. At the same time, we can't understand either the Old or the New Testament if we read them ignoring the principle of mercy. These texts are not morally or religiously "neutral."

What is the meaning of the "principle of Divine mercy" in biblical interpretation? How should we approach the Bible?

Before we open this Book, we need to acknowledge that the Word of God is designed to perfect us, our souls, minds, and lives. Only then can we be led by it instead of trying the opposite—to perfect ourselves with our little reason. I have an impression that many scholars of today do not appreciate Holy Scripture enough. They naively think that they have already understood and plumbed the depths of God himself. Meanwhile, the words of the Bible remain beyond our comprehension; they are more perfect than we can imagine. Ultimately—perhaps only in the future life—we will realize that God, using more or less fallible people, conveyed His truth infallibly without a trace of any error. It is only our smallness and sinfulness that make us unable to see it.

When we open the Bible we find the narratives recounting the formation of the universe, species, and human beings. In a way, Genesis is the foundation of the Christian faith and worldview. If God did not make all of this, then the history of salvation hinges upon a void, it turns into a merely humanist myth or, as Dawkins observed, it resembles the primitive stories of ancient tribes. When we open works by contemporary biblical scholars of any denomination, they usually begin with a series of reservations, such as cautioning us that we shouldn't approach this text literally, that it is just a metaphor. Does this mean that Genesis should not be treated seriously? How should we understand the Genesis account of creation when contrasted with ancient mythologies?

A fragment from a book by George Weigel comes to my mind. Let me quote it here:

What many Catholics in the West have learned from modern biblical scholarship is a profound distrust of the Bible: this didn't happen; that's just a metaphor; this is a myth. The Bible that was to have been loved, and that was to become a means for encountering the living God and his Son, has become in too many instances an artifact to be dissected, leaving unlovely and even repellant remains on the dissecting-room floor.[1]

What Weigel says is unfortunately true. When you look up contemporary commentaries on Genesis, right from the beginning you would learn that this text must be approached—if not as just a regular fairy-tale—then at least with a high dose of suspicion. It definitely does not tell a true story. It is claimed that, for instance, you should not follow what stands literally but rather decipher what the inspired Author intended to say. This thinking is inspired by the conviction that the Bible, by itself, can only deceive us, that without our human rational effort and advanced exegetical tools the Scriptures do not convey any understandable message. This thinking, however, goes against the more fundamental Christian conviction that God gave us Holy Scripture not to plunge us into interpretative difficulties, but rather to reveal the truth, and to do it not only to the exegetes but to all people.

Today's Bible scholars say that the main goal of the Genesis account of creation is to justify the Sabbath, or it is just an intricately woven poetic text, or merely a reminiscence of pagan mythology in Hebrew culture. Each of these interpretations may reveal some aspect of Genesis, but claiming that Genesis does not contain any other, more fundamental truth is a profoundly reductive approach to the holy text.

Does that mean that theologians sometimes diminish or ignore the message of the Bible?

A poster-child claim of modern theologians "concerned" about the "proper" understanding of the Bible is their saying that Genesis is not a history or biology textbook. But even Young Earth Creationists typically do not say so. Let's consider how far this thesis can take us. Does the fact that the Holy Scripture is not a textbook in the modern sense justify our dismissal of its content? There are many instances where we get to know historical or scientific truths not from the history or science textbooks. When a father explains to his son the design of the combustion engine, he does not use an engineering textbook, yet what he says is fact, not fiction. When a grandma recounts to her grandchildren events remembered from World War II, she does not create a history textbook, but neither does she tell fake or merely figurative stories. Indeed, she speaks about true events that took place in a specific time and place. Similarly, from the fact that the Holy Scripture is not a history textbook it does not follow that it does not convey true history. From the fact that it is not a biology textbook it does not follow that it does not say anything about nature or the way that the universe was formed.

Does this mean that Genesis is not a natural history textbook, but it can be used to fulfill a similar task in our knowledge?

Not exactly. The question about the origins (the origin of the universe, the origin of species) is not a scientific question. That's why in answering these questions your primary source is not a science textbook, but Genesis.

Yet, science seems to deal with these questions. At least this is what many scientists recognize as a scientific domain.

Only after Darwin and not all of them. Copernicus, Newton, Kepler, Galileo, and even Lyell, Cuvier, and many other distinguished scientists knew the limits of science and did not trespass them. Newton, for instance, explicitly taught that the planetary system functions owing to the laws of physics, but the same laws could not have put the planets in their orbits and create the system in the first place. Similarly, we could say that Darwin explained how natural selection can produce some variation among organisms, but not how all species and life itself originated. Surely, it was an irresistible temptation for the scientists to enter and occupy the field of origins, and it will be difficult for them to withdraw and give it back to its proper domain of theology. But the "law of acquisitive prescription" does not apply here. If science has limits, they remain the limits of science, regardless of whether scientists respect them or not.

Is there any theological argument for the limits of natural science?

First, we need to ask what is the sense of supernatural revelation in general. Would God enlighten the biblical authors to reveal to humanity the history of the beginnings if all of this could be discovered by natural investigations? If we believe that God is the first Author of both Scripture and nature, then it is reasonable to accept that He communicated to us one coherent story of the origins rather than two contradictory messages. If the evolutionary story were true, why would the Bible tell something entirely different? Why wouldn't it give any hint pointing to evolution rather than creation?

Such a vision of Divine revelation does not fit anything else we know about God's pedagogy. Instead, I think that God communicated through the Bible those truths that we cannot learn from the book of nature—those that are beyond the capacity of any natural cognitive effort. The two books do not contradict each other and they barely overlap. Instead, they explain and supplement each other. One of the fundamental truths contained in Genesis is, for example, that man was formed from dust, not that he evolved from the population of some mythical hominids. Science cannot overturn this truth of faith, because even if the so-called "hominids" (half-ape, half-man creatures) had ever existed, they would not have constituted a proof against the special creation of man.

The book of nature and the book of revelation supplement each other, but the more important truths are the ones revealed in the Scriptures precisely because we know them only by the authority of God. They concern the ultimate things—the ultimate beginnings and the ultimate destination of humanity and the universe. They are also more certain than the truths discovered by natural investigations, because human reason is fallible while God is infallible. This is why the proper Christian attitude is to trust in the Bible: Simply believe in the word of God, even if in a given time and circumstances it seems incomprehensible to us.

Are there any other ways of explaining away the Holy Scripture?

Another interesting way of invalidating the Bible is saying that Genesis does not tell us anything about *how* the universe was made, but only *that* it was made. This idea is captured by the catchy phrase attributed to Galileo: "The Bible tells us how one goes to Heaven, not how the heavens go." Again, we see here reductionism—this time the metaphysical and historical contents are voided and the biblical message is reduced to the moral teaching alone. According to this approach, it is not the task of the Bible to create a worldview in our minds, but only to communicate the moral norms—"how to get to Heaven."

This is a very dangerous instance of reductionism, because *how* one should act heavily depends on *who* or *what* one is. If we do not know the latter, we can hardly know the former. At the same time we need to notice that the historical reading of Genesis does not allow us to explain *"how the heavens go"* but rather *where* the heavens *come from*. The human cognitive effort, such as the one we observe in empirical sciences, may lead us to amazing results in answering the question of *how* the universe works, but the same effort will tell us very little about *where* the universe came from. This is why the Bible gives us a worldview, that is, it explains how the universe originated and where it is ultimately headed.

If we did not have the biblical revelation, we could think, for instance, that all the diversity of creatures came about through the emanation of one type of being from the other, or that the universe is eternal. This was the thinking of the pre-Christian pagans, and this is also how today's evolutionists try to explain origins. The evolutionary cosmogony constitutes the bases for the post-Christian worldview dominating present culture. But thanks to the Bible we know that it is not true. We know that it was God alone who formed the essential elements of the universe by His supernatural and direct power that He used over a definite time that the Bible calls *the six days*.

Let's return to the main question: How should we understand Genesis?

The account of creation should be understood the same way that the entire Church tradition has understood it, that is, as a literal and historical truth. And this is what differentiates the Bible from the myths—myths are made up, which means they speak about a past that was never present, whereas the Bible speaks about the past that at some moment was present.

This approach does not necessarily mean that the universe is just six thousand years old, or that dinosaurs existed along with humans. The length of the biblical days of creation has been disputed among both scientists and theologians since Antiquity. As early as the 5th century, St. Augustine stated that the day from the Genesis narrative could not be understood as a natural day (12 hours). Following the diversity of the opinions among Church Fathers and holy Doctors, the Catholic Church confirmed in 1909 that the Hebrew word *yom* standing for "day" in Genesis, chapter 1, can be understood either as a natural day or as any other period of time. Therefore, there is no contradiction between the literal understanding of the Bible and the origin of the universe dating billions of years ago.

If we look at the data delivered by modern sciences, such as paleontology or geology, we see that they do not contradict the literal reading of the Bible. The general framework of the Genesis account and of the story presented by modern cosmology is the same. In contrast, Darwin's theory is an ideological overlay on reliable data that is designed to explain away the supernatural activity of God from the history of the formation of the universe. We can therefore say that the Bible is compatible with the scientific data and facts while it contradicts Darwin's theory. Nevertheless, we need to distinguish the literal and historical reading of Genesis from "Biblicism" or something that we could call "biblical materialism."

What is Biblicism?

In the traditional approach to the Bible the exegete's task was to establish what the literal sense of particular passages and words were. This is why Augustine and other Church Fathers wrote the *literal* commentaries on Genesis. In contrast, Biblicism assumes that there is just one literal sense which is identical with the most common meaning of the words. Based on this assumption, "biblical materialists" exclude any other understanding of the word "day" from the creation account than the natural day. Another example is the first verse of Genesis: "In the beginning God created the heavens and the earth." For the biblical materialists "the heavens" means the sky (earthly atmosphere) and "the earth" means our planet. In contrast, there is an established Christian tradition interpreting "the heavens" as the spiritual world and "the earth" as the material world. And this is still the literal approach to the text.

Why do theologians deprive the Bible of its meaning?

I think the first reason is the modern separation of Scripture from the Holy Tradition. For centuries the Catholic Church accepted Tradition as the interpretative norm for the Bible. If there were doubts regarding the understanding of a given passage, the answer was sought for in the writings of the Church Fathers and the medieval Doctors. If they substantially differed on a given issue, then the plurality of possible solutions was accepted. Besides, it was commonly accepted that biblical texts have a few parallel senses (traditionally four of them were distinguished—the literal, the allegorical, the moral, and the anagogical) but the literal sense was considered the basic one, in the light of which all other senses should be judged. Thus, the patristic and medieval approach to the Bible was not reductive or closed to multifaceted approaches and interpretations.

When Martin Luther came up with his principle of *sola Scriptura* (the Scripture alone, without Tradition), the Bible was isolated from its proper interpretative milieu. And because the text of the Bible is not easy, and by its very nature it demands some interpretation, there was the need to come up with some type of interpretation other than the golden standard of Tradition. This is how Luther and his followers invited a new approach called "critical exegesis." Critical exegesis is a human effort based on different scholarly methods that is designed to discover the "true" meaning of the Bible. To be an exegete in this sense one does not need to believe but needs to be educated. One does not even need to know the teachings of the Church, the dogmas of faith, but has to be acquainted with the theories and methods that are applied to the ancient texts. Ultimately, critical exegesis leads to the odd conclusion that a non-believer, an atheistic scholar, may be considered a "better" interpreter of Holy Scripture than a saint or a Doctor of the Church.

Is critical exegesis a problem?

By applying critical exegesis, it is very easy to deprive the holy text of its essential meaning. In the critical perspective the Bible is not the norm of our thinking, rather the scientific method becomes the norm of interpreting the Bible. This goes against the traditional approach set by Augustine, who says that the Holy Scripture is *norma normans non-normata* (from Latin: "a norm determining [other] norms, undetermined by [any other] norms"). By this Augustine means that there isn't any method or pre-knowledge, or a system of thought, that could determine our understanding of Holy Writ. It is the opposite—any method or any presupposition is judged, "normed," measured, by the divine word of Holy Scripture itself.

The critical approach perfectly suited the positivist mentality of the nineteenth century. On the one hand critical exegesis was "scientific" and, on the other, it helped to invalidate inconvenient fragments of the Scripture in such a way as to make them compatible with different nineteenth-century "-isms," among which Darwinism occupied a prominent position. Some of the more zealous Protestants wanted to counter critical exegesis, but for them the only alternative was to stick to the letter as closely as possible. And this is how biblical fundamentalism was born. In the context of Genesis, biblical fundamentalists assume not just separate creation of species, but also the short history of the universe. They think that if you allow an old universe, you need to stray from the text and then everything is permissible. Hence, among Protestants there is a division between scholars who invalidate the Scripture using critical exegesis and fundamentalists who dismiss scientific evidence.

Did the developments you described have any impact on Catholics?

Unfortunately, Catholic scholars promptly accepted the methods of critical exegesis. In the first phase it caused a serious conflict in the Church called the "modernist crisis," which saw its first opening at the dawn of the twentieth century. Over time, however, those methods disseminated throughout Catholic scholarship and—for the most part —they supplanted the Holy Tradition. Today, even though Catholic theologians are formally bound by the decrees of the Council of Trent, they *de facto* work according to the principle of *sola Scriptura*. It is enough to say that an average Catholic seminarian can finish the whole theological curriculum without reading a single patristic or scholastic commentary on a book of the Bible. Yet, the same seminarian will spend a lot of time studying many various, contradictory, unproven, and oftentimes methodologically incorrect theories of the origins of the biblical books.

In the context of the biblical account of creation, theologians often deprive it of the historical meaning, because they think that if the biblical story was historically true, they would have to accept the young age of the universe. This, however, would contradict scientific data. In order to avoid the so-called "science and faith conflict" they prefer to claim that Genesis is nothing but an intricately woven religious poetry. But had they known the Tradition and adhered to it, they would know that holy Fathers and Doctors did not have a common opinion regarding the length of the days of creation.

In contrast, the immediate and supernatural formation of the first human body was always considered a truth of faith. Therefore, there are no obstacles from the Tradition to acknowledge the scientific concept of "deep time." At the same time, however, the biblical and theological argument against the evolutionary origin of man is overwhelming. In my opinion Catholic theology will never regain the proper understanding of the Bible, including the answers to the questions of origins, unless Tradition is restored to its proper place in biblical scholarship.

> *And what about the Galileo affair? Holy Doctors, Church Fathers, and many saints firmly believed that, for example, mountains were formed supernaturally and immediately by God. Today, we know that they were formed naturally. Similarly, they all accepted the geocentric model of the solar system, which became the bone of contention in the argument with Galileo. Is that not evidence for the fallibility of Tradition in the interpretation of the Bible? If so, how could we still stick to the literal interpretation?*

First, we need to notice that the geocentric model of the cosmos was not a Christian view created under the influence of the Bible. Rather, Christianity, along with the entire civilization of Antiquity, adopted the earlier cosmological models created by pagan astronomers, such as Ptolemy. *Geocentrism* was accepted by educated Christians not because it stemmed from the Bible, but because it simply did not contradict it. But *heliocentrism* does not contradict the Bible either, because neither of the models is the proper object of the biblical message. The Bible simply does not explain the workings of the universe on the physical and scientific levels.

> *Does it mean that science cannot modify our understanding of the Bible in any way?*

Surely, one could say that if science proves something beyond reasonable doubt, then faith must be modified according to the new discovery. As an example of such modification the heliocentric theory of Copernicus is usually brought up. All Christianity until the sixteenth century believed that the Earth is an immobile center of the universe. This view was present in Church Fathers and medieval scholastics. This cosmological exceptionality of the Earth nicely fit the theological truths about the distinguished place of man in nature and the exclusiveness of the Incarnation. But this worldview fell apart along with the Copernican revolution, that is, the discovery of the Earth's orbital motion, and later a more and more detailed picture of the universe.

Many contemporary scholars believe that a similar process happened due to the "Darwinian revolution." They think that universal common ancestry has been proved beyond any doubt and thus invalidated the belief in separate creation of species. Moreover, atheists and materialists used Galileo's case to intimidate those theologians who would dare to say "stop" to the expansion of naturalism. In effect, theologians of the first half of the twentieth century reasoned along the following lines: "Science and faith cannot contradict each other. Therefore, if scientists say that man, regarding his body, was not created, but evolved in the natural process of development of "living matter," then we need to figure out how to modify our understanding of the Bible in order to make it compatible with the evolutionary idea. This way we will cleverly anticipate the facts and avoid a second Galileo affair." This kind of thinking overwhelmingly won over the Catholic intellectual community during the twentieth century.

Was this reasoning correct?

This reasoning contains some unfounded assumptions and also some confusion in concepts. I will point out just two major problems. First, if it comes to the conflict between faith and science, why should we assume that the religious concept (the belief in creation) is wrong? Maybe this time people who speak in the name of science made a mistake. Maybe they wrongly interpreted data, or extrapolated limited observations beyond the scope of science. In the classical understanding of faith and science, it is faith that provides certain knowledge whereas natural science is changeable and fallible. Thus, we should trust a theological concept as long as nothing contrary to it is proven beyond reasonable doubt. And this is the case with biological macroevolution or universal common ancestry, or the simian origin of man. These are all abstract claims far from being scientifically demonstrated, which are even rejected by many scientists.

But the second problem is even more important. It stems from the fact that there is no direct analogy between Darwin and Copernicus. These two authors offered two different theories to answer two completely different questions. Generally speaking, Copernicus's theory gives an answer to the questions: How was the universe built? How do the planets move? What kind of relations and forces govern them? etc. And these are strictly scientific questions, because they concern the operations of the universe. In contrast, Darwin attempted to answer an essentially different question, namely, Where did life in its different forms come from? Again we return to the key difference between the two types of questions: *how?* and *from where?* The first is properly addressed by science, but the second goes beyond its scope. The answer to the second question needs to involve special revelation, which we find in the Bible.

Do the evolutionists respect the limits of science?

No. The confusion regarding these two types of questions is seen in their classic argumentation. They explain, for example, that malaria parasites through genetic mutations and natural selection can acquire resistance to drugs and they present it as evidence for how the Darwinian mechanism works in practice. No creationist, even the most fundamentalist, has problems with this demonstration. These are facts. But evolutionists do not stop there. They claim that this same mechanism that accounts for the minor adaptations can explain where malaria parasites came from in the first place, or how life itself began.

We see therefore an unfounded extrapolation of some biological mechanism from micro- to macroevolution and a similarly unfounded transition from the question of *how?* (how the parasite deals with an antibiotic) to the question of *where from?* (where the given species of creatures came from). In other words, it is claimed that the same mechanism that causes adaptations within species may explain the origin of species. This is wishful thinking with no support in scientific evidence.

As evolution critics have observed, the survival of the fittest does not explain their arrival. Since malaria parasites deliver the best examples of Darwinian adaptations (due to the enormous populations and fast succession of generations) and they do not change into anything else, this shows that the Darwinian mechanism cannot really do much. In particular, it cannot create anything fundamentally new, for example a new form of life.

Perhaps what the Darwinian mechanism cannot achieve in a short time becomes achievable in a long time. At least, this is how Darwin's proponents try to overcome the problem.

We need to realize that time is not a cause here. And the absence of a cause remains the absence of a cause regardless of whether it lasts a year or a billion years.

Does all this mean that we shouldn't fear another "Galileo affair"?

The progress of science not only delivers us more and more answers to scientific questions, it also helps us see the very limits of science. Today we already have a clear criterion for distinguishing why Copernicus and Galileo could have been right and Darwin couldn't. It is because the former gave a scientific answer to a scientific question while the latter attempted to find a scientific answer to an essentially religious question —the question about the origin of the diversity of living beings. The very shift from the theological level of discussion to the biological level made Darwin's theory intrinsically reductionist. In my opinion the fear in some ecclesiastical circles that a new "Galileo affair" might happen if the Church condemns Darwin is fueled by an incorrect understanding of the problem, specifically of the limits of science and the limits of theology.

Someone could protest by saying that geocentrism was also taught by theologians in order to defend the traditional interpretation of Scripture. In fact, the very words of condemnation filed against Galileo in 1616 made it clear that he contradicted both philosophy (i.e. science) and Holy Scripture. Why should we accept the arguments of Galileo/Copernicus and reject the arguments of Darwin?

I have already given one criterion for making this distinction: C,osider what kind of question the scientific theory addresses and you will know if the theory is scientific. Does it actually concern the natural events within the universe? Or maybe it is a historical theory, or maybe it is just a kind of "reductionism," that is, an attempt to break down the phenomena of a higher-order to the causal explanation of a lower order. In the case of Darwinism, we deal with reduction of God's supernatural and immediate activity to a natural process occurring according to the laws of nature. Moreover, *geocentrism*—even though it was convenient for theologians for philosophical reasons—was never taught as a truth of faith.

We need to take into account not just the question of whether the holy Doctors and Church Fathers held some concept as true based on Holy Scripture but also whether they considered it a truth of faith revealed by God. The creation of the human body from the dust of the earth was considered a revealed truth, whereas the geocentric cosmology belonged to the "secular knowledge" taken from the pagan astronomers. This is why the first cannot be overturned by science but the latter may change according to subsequent scientific theories. The Bible does not support geocentrism; at best, it is compatible with it, just as it is compatible with many other false scientific theories. It is simply not the goal of the Bible to settle the scientific issues.

When a star explodes, something new, a supernova, begins to exist. Similarly, there are many things on earth which we know arise in a natural way. In the geological record, we see mountains growing due to tectonic plate shifts (and other factors) or rivers changing

*their courses. So, evolution in nature can create new things. Thus, it
can address the questions of origins, for instance, the origin of new
white dwarf stars. Does it not invalidate your criterion based on
the two types of questions (*how *and* from where)?

To address this problem, we need to introduce one more distinction.
There is a substantial difference between the origin of a completely new
form of life and the emergence or transformation of some non-living
entity. Evidently, natural processes of changes in the universe cause things
like the emergence of supernova from perishing stars, the galaxies, or
mountains and valleys on the surface of planets. But in these types of
changes nothing fundamentally new is created.

Stars are formed by a collapse of galactic nebulas; then they reach
the state of equilibrium, burn over a few billion years, turn into red
giants and finally explode. At the end there are white dwarfs and a great
amount of interstellar dust. This process is often called the "evolution"
of stars. Evolutionists readily use this example to propagate biological
macroevolution. They say something like this: "Everything evolves: stars,
planets, geological formations, languages, and cultures; in the same
way, the biological realm also evolves." But this reasoning is misleading
because it contains a blend of ideas where the word "evolution" is used
a few times with different meanings. If we call the burning out of a
star "evolution," then the analogy in biology would be—at best—just
an ordinary succession of generations. A cat begets a cat, which is born,
grows, matures, ages, and dies. This doesn't mean that a cat will change
into some other type of creature. Similarly, a star will not change into
anything fundamentally new. "Stellar evolution" has nothing to do with
biological macroevolution except for the failed analogy.

The origin of species entails something completely new—new, coherent information—or, speaking philosophically, a completely new *form*,
or a new *substance* of a living being. And this is beyond the reach of
nature. It demands the work of an all-powerful intellect which is capable
of introducing a new design into nature that constitutes an irreducibly
complex whole. Again, Edwin Hubble and Fred Hoyle could have been

right when speaking of the expansion of the universe or evolution of stars. Charles Lyell could have been right when speaking about the emergence of new geological forms thanks to the laws of nature operating over vast periods of time. But Darwin wanted to explain the origin of functional biological novelties and entirely new species by reference to continuous material processes. This was a search for a cause where it could not have been found because of the very nature of the issue. Science can indeed help us properly interpret the biblical text concerning nature, but the philosophical usurpations of materialists are something different. Even if they dress up in scientific garments, they do not have the same value as solid science.

If this is the case, how should a believer react when he encounters theories presented by scientists which contradict faith?

The problem of using scientific theories to disprove faith is as old as Christianity. Already in the first centuries Gnostics and different pagan philosophers claimed that the faith presented in Holy Scripture was untenable in the light of what we knew about nature. Today, from the perspective of the two millennia that have passed, those serious "scientific theories" that supposedly disproved Christianity and the Bible make you smile or feel pity in the light of current knowledge. Yet, pagans of the first centuries were just as firmly convinced of the veracity of their theories as modern evolutionists are convinced of biological macroevolution.

We know that contemporary science analyzes problems on a much deeper level and gets to the very bottom of material phenomena. This seems to give it more credibility than ancient science based on abstract principles or very general observations. Modern science is independently confirmed by its technological success. Can ancient science really compare with modern science?

I would agree with this thesis, but only within the strictly experimental sciences. Darwin's theory, understood as biological macroevolution, does not belong to such strictly defined scientific categories. In this sense it is not, as I already said, a scientific theory, but rather a certain idea shaping people's worldview today. Long ago St. Augustine explained how Christians should proceed when faced with a conflict between natural knowledge and Holy Scripture. Even though Augustine's advice was given in the pre-scientific era it remains legitimate, because it was repeated by Leo XIII in the encyclical *Providentissimus Deus* [The God of All Providence] in the late nineteenth century, at a time when positivism and naturalism were gaining a predominant position in Western culture. In this way Augustine's response was made a part of Catholic teaching.

And what, exactly, did St. Augustine advise?

Augustine taught that when it comes to an apparent conflict between faith and science, the first step is to show that the theory of nature assumed by modern science is compatible with Holy Scripture. If this is not possible, then Christians should show that the modern theory of nature is false as far as it contradicts the faith. If this turns out to be impossible, Christians should keep believing "without any shadow of a doubt" that the modern theory of nature is false. We see that Augustine was aware of the fact that sometimes materialists and naturalists abuse science in order to fight religion. Augustine also knew that Christians may not immediately propose good arguments to fend off their claims. Even so, he gave priority to faith, assuming that the truths of faith are beyond the reach of science. Science can neither invalidate nor modify them.

Did the theologians of the twentieth century follow the advice of St. Augustine and Leo XIII?

No, not at all. Augustine's *modus operandi* consisted of three steps. First, show that the scientific theory is compatible with Holy Scripture. Contemporary theologians say something like this: "There is no contradiction between Holy Scripture and the evolutionary origin of man.

Therefore, there is no problem with believing both in the Bible and in Darwin's theory." But, as we already observed, this alleged compatibility has been achieved at the cost of voiding the biblical account with the help of critical exegesis and the rejection of the holy tradition. Thus, it cannot be a Catholic position.

But there is also another problem connected with this thinking. Apparently, the theologians who say so do not understand, or they do not want to understand, what evolution is and what creation is. Generally speaking, creation is a direct and supernatural calling of being into existence. In contrast, evolution is a natural process operating in nature, which theistic evolutionists call a secondary cause (an instrumental cause) of creation. But the same process cannot be at the same time natural and supernatural, direct and mediated. This is why, logically, evolution and creation are mutually exclusive: either one or the other. Darwin himself understood it quite well when he wrote that you can either believe in the struggle for life and natural selection or in a multitude of "repetitive" acts of creation. Theologians who pronounce the "compatibility of evolution and creation" actually distort the notion of creation, or evolution, or both. For instance, a renowned Catholic philosopher of the present day who is deeply convinced that evolution does not contradict creation, claims that creation does not mean "the beginning of things" but rather "their dependence in being on God." Well, if we remodel the classic notion of creation to the degree that it actually means something different, then we can say there is no contradiction. Except that we are not Christians anymore.

Other scholars believe that evolution is a natural process in which God is constantly present and somehow, "from the inside," He guides the process to its end. Due to the works of a few theistic evolutionists, such as Arthur Peacocke, John F. Haught, John Polkinghorne, and others, this is the most popular version of theistic evolution in both Catholic and Protestant circles. Leaving aside the pantheistic provenience of this concept, it is quite clear that no scientific evolutionist understands evolution this way. For naturalists, such as Darwin and those who created and popularized the neo-Darwinian synthesis, evolution is a purely natural, blind, and aimless process about which we are regularly reminded by the guardians of evolutionary orthodoxy.

Let me bring up just two characteristic passages. In the 1970s one of the greatest protagonists of evolution, the French molecular biologist Jacques Monod, assured his readers

> . . . that chance alone is at the source of every innovation, and of all creation in the biosphere. Pure chance, absolutely free but blind, at the very root of the stupendous edifice of evolution: this central concept of modern biology is no longer one among many other possible or even conceivable hypotheses. It is today the sole conceivable hypothesis, the only one that squares with observed and tested fact.[2]

In the same tenor but more recently, Richard Dawkins affirmed that "the universe that we observe has precisely the properties we should expect if there is, at bottom, no design, no purpose, no evil, no good, nothing but pitiless indifference."[3] We see therefore that theologians who want to squeeze in God into evolution stray from the scientific understanding of this process. In their case the alleged compatibility between evolution and creation is obtained at the cost of the substantial redefinition of evolution.

And what about the remaining steps of the Augustine's advice?

The second step in Augustine's "test" requires showing that a theory of nature is false on natural grounds if it contradicts the faith. Very few scientists moved on to this stage. Among the honorable exceptions I should mention a late Jesuit priest and anthropologist, Fr. Peter Lenartowicz, who based on solid philosophical grounds, with great academic excellence, demonstrated the biological problems of the human evolution from apes. Among the currently active there are scientists connected with Discovery Institute, Michael Behe, William Dembski, Stephen Meyer, Rick Sternberg, Douglas Axe, and others. These are accomplished scientists, with PhDs from top ranking universities, and none of them is a Young Earth Creationist or biblical fundamentalist. They came to questioning Darwin by thorough study of the evidence and a kind of "openness of the mind" which should be always welcomed in the scientific community. In some cases this "openness of the mind" costs a scientist an academic career or a high-profile job. Today we have come to the point in our society that questioning Darwinian ideology requires not just broad professional knowledge but most of all courage. It's a pity that so few theologians possess both of these attributes to join the scientists.

Let's go back to Augustine. The "Augustine Test" is a sort of instruction manual on how a Christian should proceed when he encounters a theory opposed to faith presented by "science." What is the third step of the "test"?

If it were not possible to refute a scientific theory that contradicts faith, Augustine still recommends believing without the slightest hesitation that the theory is false. Augustine was confident that a Christian should adhere to the faith, because he knew that science and faith have one ultimate source, who is God. And God does not contradict himself.

So, if we followed the advice of the great Saint, even if we had no scientific arguments against the evolutionary origin of man, we would still believe in the creation of the first human body?

Most modern theologians when facing the pressure of the so-called "scientific community," which is always reflective of Darwinian propaganda, shy away from classic Christian positions. They accept the evolutionary origin of man, because they believe that science requires it. However, these same theologians usually believe in the resurrection of Christ or the Virginal Birth of Jesus. And yet these truths, if judged by the same standards, turn out to be deeply unscientific. Scientists can present massive evidence that virgins do not give birth or that dead people do not come back to life by themselves. Nevertheless, Christians firmly believe that there were events in the past that transcended the order of nature and although they contradict the naturalistic understanding of the universe they are not contrary to reason.

Hence, I do not understand why it is not allowed to believe that man was formed from the dust of the earth. Why should we accept the miracles of the history of salvation and at the same time deny the miracles of the history of creation? Theologians who adopt this attitude have failed Augustine's test. They think they follow "science", but in fact they undermine the faith and at the same time they ignore or a priori dismiss the serious scientific arguments against the simian origin of the human body. By adopting such attitude, instead of building a synthesis of science and faith, they actually dismiss true faith and ignore real scientific evidence.

3. Of Apes and Men

Let's get straight to the point: Did men descend from apes?

You are asking about the central issue of the entire debate on evolution. The origin of species would not be so controversial if it were not for the conclusion that if all species share common ancestry, then men descended from apes. Of course, many contemporary evolutionists are more willing to talk about a common ancestor of men and apes, as if this this formulation were less radical and made the very idea somehow more socially acceptable. But note that changing the formula from "men descended from apes" to "men and apes have a common ancestor" is just wordplay, because the common ancestor would be at best an ape.

There is a biological chasm between man and ape at the level of genus. However, evolutionists try to hide this fact by creating so-called "intermediate forms" or "missing links." Simplifying the matter, we can say that intermediate links are produced more or less like this: A paleontologist finds a piece of a bone, or a tooth, or a few scattered fragments that do not match anything he has seen before. Guided by the assumption that man and ape share common ancestry, he concludes that these are the remains of another "hominid," that is, a creature which is half-man, half-ape. In this way, a fairly large group of hominids has been created, and the average non-scientist has the impression that there is a whole lot of evidence for the existence of these mysterious species.

But there is very little empirical material to support this thesis. In fact, the whole concept is based on the assumption that such creatures must have existed. If they did not, the leap from ape to man would be beyond the reach of natural evolution. Notice what is going on here: The poor and very ambiguous empirical material is forced into the framework of the theory and becomes evidence in support of it. Fossils are interpreted as intermediate links because that is what the theory requires. And how do we know that the theory holds water? Because the fossils interpreted in this way support it! This is typical circular reasoning, by which the conclusions are embedded in the assumptions. There is no way to argue against such "logic."

So in what ways are we similar to apes?

Are we? When I look at an ape, the first thing that catches my eye is the huge differences. It is no coincidence that calling a human an ape is considered offensive. But seriously, in nature everything is similar to everything else. In other words, whatever element of the visible universe you consider, you will always find similarities with other elements of this universe. Take, for example, a human being and an ordinary fork that you use in the kitchen. The differences are obvious: A fork is typically made of more than 90% of one metal or another. A human, on the other hand, is made mostly of oxygen (65%), carbon (18.5%), and hydrogen (9.5%), in addition to a half dozen other elements present in smaller amounts.[1] But you will find some carbon in a fork, just as you will find some metal (such as iron) in a human body. So, there is some similarity. However, it does not follow that a man and a fork share common ancestry the way evolutionary biology understands it. Similarity does not translate into kinship.

Yes, but in the case of man and ape we are dealing with organisms that live and reproduce. Kinship is therefore possible.

Nevertheless, the principle that kinship, or common ancestry, cannot be inferred solely on the basis of similarity remains valid. You need to assume common ancestry to infer it on the basis of similarities. And if you do not assume it, then you may conclude a lack of common ancestry even if the things are very similar. To realize the problem, you may also ask what kind of dissimilarity would lead evolutionists to a conclusion that there is no kinship? There is no such option. If two specimens are not similar, they say there is a gap in the fossil record rather than that they do not have a common ancestor.

Are you saying the claim of 99% genetic similarity between men and chimps is irrelevant, then?

Glossy magazines, popular portals, and other non-scientific sources regularly repeat this myth. By the way, any person you ask will give you a different number. However, these calculations were made many years ago, when we did not yet know the entire human and chimpanzee genomes. These calculations did not take into account the regulatory regions that constitute the majority of the genomes, but only the coding regions. Recent research suggests a similarity between the genomes of men and chimps of about 88%. Some geneticists speak of 76%. But, again, for the details of these studies, regarding, for example, the methods of comparing gene sequences, I refer you to the professional literature.

If that is the case, has the much lower percentage of similarity between men and apes caused evolutionists to back off the assertion that similarity proves common ancestry?

Perhaps, but we need to look at the problem from a broader perspective. First of all, according to neo-Darwinian logic it is assumed that a living organism is primarily its genome. This is a reductionist approach that is rejected by an increasing number of biologists today. In the past, most scholars thought something like this: "Give me a genome and I will create a new species." Today, we know that genes do not contain all the information needed for the functioning of an organism. Therefore, no matter how many changes, how many mutations, occur in the genome, we will not obtain a new species. And for this reason alone, neo-Darwinism is dead.

Today, the DNA is increasingly viewed as information that an organism uses in its functioning. British biologist Denis Noble even called the DNA a "sensitive organ" of the organism. In any case, it is clear that genes are not humans and genes are not apes. Therefore, even a complete identity of the human and chimp genomes would not be enough to make them the same species. Perhaps a good analogy between the genetic material and an organism is a computer and a flash drive. On the flash drive we have information—a program that the computer reads. In order for the computer to read the flash drive, it must have all the the information—both the machinery (hardware) and the programming (software)—that existed before the flash drive was inserted. Claiming that changing the genetic information will turn a chimpanzee into a human is like saying that inserting a new flash drive or a CD into a computer will turn it into a TV set or a kitchen appliance. Of course, a computer can be used to watch movies, but that does not mean that the computer becomes a TV set. Similarly, a human can climb trees, and an ape can take a hammer and pound a nail with it, but that does not mean that the ape becomes a human.

Scientists have repeatedly announced the discovery of missing links between humans and the great apes. Doesn't that make the gap between humans and monkeys very narrow, even if the genetic differences are more significant?

As I said, the existence of many intermediate links is an assumption, not a conclusion from research. Therefore, whatever scientists find in the fossil record will be classified as an intermediate form. Let's imagine that one day paleontologists found a fossil of an elf, Pegasus, or a Sphinx that lived on Earth millions of years ago. Do you think that this would create any difficulty for the theory of evolution? No, because you can always find some similarities between these organisms and those currently extant. Since they are similar, they are the ancestors of the current forms, according to evolutionary logic. This way, anything can be included in Darwin's tree of life, even such absurd links as those cited here.

Any finding can be an element of the tree of life, because in nature everything is somehow similar to everything else. If the similarity is barely noticeable, evolutionists can always say that either we do not know the missing links yet, or that this is a completely new "evolutionary path" that we did not know about until now. I prefer not to judge how much solid science there is in this type of inference.

The problem is that exactly the same type of reasoning applies to those mythical hominids. Scientists critical of Darwinism easily point out that all fossils of primates can be classified as either ape-like or human-like. In many cases, the remains are so incomplete that their classification depends solely on interpretation. If, therefore, a scientist approaches the evidence with the assumption that common ancestry is the only explanation, then each new fossil becomes a new hominid. And that is why the number of hominid species is huge and constantly growing. Often, a new species of a hominid is based on just one fragmentary fossil. It even happens that one fossil receives several names and in popular perception functions as several different hominid species. When we delve into this mess, one may even suspect that there is some purpose to it. As if the large number of most bizarre names was intended to convey to the general public one simple message, such as: "There is no doubt that man descended from apes, because there is an overwhelming number of intermediate links."

Is the existence of intermediate forms evidence of the common origin of man and ape?

I don't think so. Most of all, however, I do not believe that there are any intermediate forms. If someone claims that a short man with black skin is a missing link between man and ape, this only proves how deceptive evolutionary logic can be. There have been many races of humans and many races of apes. The apes that we know today (about 200 species) constitute only 10% of the apes that existed in the past. If the currently known human races constitute a similar percentage of the different human races that existed in the past, then we can imagine how rich the fossil material may be, based on which many more intermediate forms can be created. But the fact that someone has a larger or smaller head, a large-boned or small-boned stature, a taller or shorter height, smaller or larger teeth, does not make him an ape or a missing link.

Does this mean that subjective interpretation plays a key role in classifying hominids?

Look how it works in the case of the forms that we know quite well thanks to the large number of fossils. Neanderthals are commonly referred to, in modern terminology, as "hominins" (direct human ancestors), that is, missing links between humans and apes (which are collectively called "hominids"). And yet we know that Neanderthals created culture, had religion, pursued arts and crafts, built houses, buried their dead, and took care of their elderly. So, why are they denied the title of "simply humans"? The only justification is the requirements of the theory, or rather the ideology, of evolution. Neanderthals date back half a million years. But if normal humans inhabited the earth half a million years ago, then none of the later forms could be called a link between humans and apes in the Darwinian sense. In this way, a large number of proposed "missing links" would be explained away. But think, if the same applies to all those supposed species of the genus *Homo*—such as *H. habilis, H. rudolfensis, H. ergaster, H. erectus, H. gautengensis, H. floresiensis, H. antecessor, H. heidelbergensis, H. rhodesiensis, H. helmei, H. cepranensis, H. longi* ("*Denisovans*"), etc., etc.—then these hominins conceived as evolutionary links exist only in the imagination of paleontologists. In reality, we are dealing with different races of people who differed from each other— just like modern people do—by their different physical characteristics or cultural achievements. But they did not differ in their humanity.

Many scientists believe, however, that the fossil material we have, although imperfect, allows us to conclude that we share a common origin with apes. Is this a correct conviction?

Even if the empirical material were as perfect as evolutionists might dream of, it would still not force us to conclude that we share common ancestry with apes. The existence of different biological forms over time proves only what it proves, i.e., the existence of different forms over time. In order to conclude that one form evolved from another, much more would have to be shown beyond the sheer fact of their existence. Biologist Douglas Axe accurately formulates this problem in an essay in the book, *Science and Human Origins*.[2] In scientific matters I give the floor to scientists. So let me quote the relevant passage:

> The truth is that humans have a tendency to accept what they've been told over and over, and scientists (being human) are no exception to this. Stories have their place in science, in the framing of ideas, but they aren't what makes good science so persuasive. So, scientists who insist that Darwin got our story, the human story, right would do well to ponder the evidence that would be needed to make that claim persuasive.
>
> Have they thought seriously about what an ape-to-human transition would entail? Have they figured out how to wire a brain for speech, or for the intelligence needed to make use of speech? Do they know how to configure lips, the tongue, and the vocal tract in order for speech to be physically possible? Have they discovered how to coordinate these inventions with all the changes needed for females to give birth to big-brained offspring?
>
> And if they've mastered all these points while wearing their bioengineers' hats, have they switched to their geneticists' hats and identified a series of single mutations that would orchestrate this whole inventive process? They may think they know some of the answers to these problems, and that's a start, but have they gone into the primate lab and done the work that should

convince those of us who wonder whether they have it right? Have they been hard at work for decades, quietly validating their ideas by producing talking chimps?

If so, have they done the experiments to measure the fitness effect of each single mutation along the line of chimps that eventually produced the ones that talk? Did they verify that each increases the fitness enough to become established in a natural population? And assuming they have checked all the boxes to this point, did they do the math to verify that the whole transition can happen naturally in an ape population within a few hundred thousand generations?

Hard questions are humbling, and humility may be the best way for scientists to earn the trust of their benefactors (the public) on this subject. In truth, almost nothing on the above checklist is technically feasible at present, so we don't need to lose any sleep over the ethical issues. My point is simply that virtually *everything* that would need to be done to establish the sheer physical possibility of turning apes into humans remains undone.[3]

In that case, I will ask directly: If man did not come from the ape, then how did he come into existence?

First, let us consider whether sciences such as paleontology and genetics are actually capable of providing an answer to this question. If man came into existence naturally, then, yes, because science explains the workings of nature. But if man came into existence supernaturally, that is, by the action of divine power exceeding the power of nature, then science cannot explain the beginning of man. Therefore, whether science can provide an answer to this question does not depend on scientific research, but on the previously adopted religious or philosophical assumptions.

So, science could solve the riddle of human origins, if we assume that this riddle is solvable by science?

Exactly. But let us go farther down this road. If we accept the biblical revelation as true, we find in it an explanation of how man came into existence. Therefore, by granting our consent to the biblical revelation, we decide to accept two things simultaneously: First, that science will not explain the origin of man, and second, that Holy Scripture tells us not only *that* man was created, but also *how* it happened. Note that the revealed message will be essentially independent of any scientific research. Scientists can continue to imagine the existence of a multitude of hominins and their evolution, and a believer can continue to believe in creation. As I said, I do not believe in the existence of intermediate beings between apes and humans, but even if I was to admit that such beings once lived, this would not invalidate my belief in the creation of the first "true" man from the dust of the earth.

Does it mean that the first man was made of clay?

Do you know of any better way for the human body to come into being? Many modern theologians not only reject personal belief in the formation of the first human body from the dust of the earth, but publicly ridicule it. And I ask, What do they have to offer in return? Stories about hominins? Man as a product of the struggle for existence and natural selection, the play of chance and necessity?

St. Augustine encountered similar reactions from pagans who ridiculed the formation of the human body from clay. He describes it in the following words:

> These enemies of the books of the Old Testament, looking at everything at fleshly, literal-minded way, and therefore always getting everything wrong, are in the habit of commenting sarcastically even on this point, that God fashioned the man from mud.

What they say is: Why did God make man from mud? Did He not have anything better, celestial material for example?[4]

In response, Augustine writes:

> What was so strange or difficult for God, even if He did make the man from the mud of this earth, about contriving a body for him that would not have been subject to decay, had the man kept God's commandment and not be willing to sin? After all, if we say that the beauty of the sky itself was made from nothing or from unformed material, because we believe its Craftsman to be all-powerful, what's so odd about the possibility of the body, which was made from any sort of mud you like?[5]

The belief in the formation of the body from the dust of the earth by the power of God was obvious to all Christians from the first century until about the early nineteenth century. There were disputes about the origin of the soul, but the creation of the body was never questioned. I believe that biology has never offered any theory that would force us to abandon the traditional faith. Today's theologians say that this faith was simply a product of an uninformed reading of the Genesis account. But as we said, they are guided by the principle of *sola Scriptura*, which means they try to interpret Scripture outside of Tradition. And that is not a Catholic approach.

For a modern Catholic, however, this may sound quite shocking.

The fact that this may be shocking does not result from the content of the truth itself, but rather from the fact that this truth has not been taught in the Church for the last seventy years or so. Besides, it is shocking mainly to learned theologians, who over the years of their studies have been taught in various ways to doubt, deny, or undermine this beautiful truth. But if you talk to simple people, they have no problem with it. I will never forget when at some closed retreat for a religious community, I was asked to talk about my research on the issue of evolution and creation. So, I talked about theistic evolution, or the so-called "New Cosmic Story," and about the traditional Catholic teaching on creation and how they differ. At the end of my lecture, someone said: "Father, we have always believed that." The point was that they had always believed in the formation of the human body from the dust of the earth. They could not understand why we should not believe it. For me, it was shocking how well the traditional Catholic belief was preserved in some communities.

So, are you proposing a return to the belief in the formation of the human body from the dust of the earth?

I postulate such a return, firstly, owing to fidelity to Tradition, but also for purely pragmatic reasons—we simply do not have any better concept of the origin of the human body. After all, we should go for the best explanation available to us. For someone educated in classical Christian philosophy, this is not difficult to understand. St. Thomas Aquinas explains this in his *Summa Theologiae*, when he writes:

> God, though He is absolutely immaterial, ... He alone can produce a form in matter, without the aid of any preceding material form....Therefore as no pre-existing body had been formed whereby another body of the same species could be generated, the first human body was of necessity made immediately by God.[6]

Currently, the alternative theories are that either man was brought to Earth by aliens, or that his body evolved from lower animals through natural change. The first concept begs the question, and the latter is impossible within Christian anthropology.

Why is the origin of the human body from animals impossible?

Because the human body is fundamentally different from the animal body. Animals have a number of features that somehow equip them to survive in nature without using reason. For example, they have shells or carapaces that protect them from predators; they have fangs and horns for fighting, sharp teeth for hunting, hair or fur for protection from the sun or cold, etc. Man, on the other hand, is deprived of these features precisely because he has reason. A pig must have a snout to dig in the ground in search for food. But man, thanks to reason, creates tools such as a shovel and a rake, and therefore does not need to have either a snout or wide claws to dig up the ground.

From a biological point of view, man is therefore the most "unadapted" animal. His locomotion is the least efficient energetically, because the two upper limbs are essentially not used. But thanks to this, man can operate tools such as a bow, a spear, or a plough. Note that if evolution were to proceed from animal to human, animal characteristics would have to be lost along the evolutionary path with reason emerging at the end. However, if the adaptive characteristics were lost before the emergence of reason, such an animal would be eliminated by natural selection as the most unfit. Therefore, we must first have reason in order to be able to lose animal characteristics. No evolutionist would claim that an animal first evolved reason.

From the philosophical standpoint, the rational soul can only unite with the human body, i.e., one that has already lost animal characteristics. In the evolutionary scenario, therefore, there is a problem of "*petitio principii*": Which comes first? The human soul or the human body? The human body cannot come first, because it would not survive without the human soul. The human soul cannot come first either, because it cannot unite with an animal body. The idea of the hominization of an animal is therefore impossible.

Do evolutionists have a solution?

Most evolutionists do not bother with anything like a soul at all. However, theistic evolutionists, those who believe that God used evolution to create man, generally look for a solution here. Many theistic evolutionists understand the beginning of man as a kind of enlightenment of the last hominin by God. The late Bishop Józef Życiński spoke of a spark of the intellectual light that was bestowed on the most "advanced" hominins. Hominins gradually took on the characteristics of the human body when, while still animals, they began to create tools, build houses, etc. On this perspective, there were a number of creatures that had technology and craftsmanship, and even art and religion, but did not have the rational soul—which means they were not human. It is obvious that the human soul in theistic evolution is understood differently from how the Catholic Church understands it. The Council of Vienne (1312), and then the Fifth Lateran Council (1517), solemnly confessed that the human soul is to be considered the only and complete form of the body (form *per se*, i.e., substantial form). However, according to theistic evolution, the so-called "hominization"—i.e., the supposed bestowing of a rational soul on the body of a hominin—is actually only an addition of a new faculty, for example; the ability to form abstract concepts or to make moral choices or to know and love God. By no means is it the creation of a new substantial form. Thus, theistic evolution solves the problem of the "*petitio principii*" in hominization in such a way that the solution contradicts the Catholic understanding of the human soul.

It seems that errors concerning the understanding of the human soul also plagued early Christianity.

The different ideas about the origin of man presented by theistic evolutionists are surprisingly similar to the old views of ancient Gnostics. St. Irenaeus, for example, describes the idea proposed by Saturianus and Basiliades, who taught that man was not formed directly by God, but by the lower angels. However, since the angels were not omnipotent, the man they formed could not stand upright but wriggled on the ground like a worm. Then, the "power from above," taking pity on him since he was created in its image, sent a spark of life, which gave the man an upright posture, integrated his joints, and made him capable of life. It seems that different contemporary proposals of how to square Christianity with evolutionary theory end up in reviving various pagan ideas, usually condemned by the Church long ago.

And what about the encyclical Humani Generis? Pius XII allowed in that encyclical the possibility of man's descending from lower animals. Was the Pope wrong?

He could have been wrong, because this statement does not meet the conditions of papal infallibility. But more importantly, Pius XII did not grant any support for the animal origin of man. He only made room for the possibility of debating the evolutionary hypothesis among Catholic scholars. One may ask why he did this, what his motivations were. In the same encyclical we will find an answer to this question. The Pope wrote that Catholic scholars should become familiar with these more or less erroneous opinions so that they could better understand them and provide a more informed response. I see an analogy here with Paul VI's later treatment of the issue of birth control. The Pope encouraged debates on this subject, not to promote contraception, but to better clarify Catholic teaching and expose the errors that lie at the root of the contraceptive mentality. Ultimately, he issued the encyclical *Humanae vitae* [On Human Life], in which he condemned contraception, despite the fact that most of the ecclesiastical experts supported it. So, we see that first a debate on an erroneous idea was allowed by the Pope, and then the Pope condemned the erroneous idea against the opinion of a majority opinion. It is possible that the same fate awaits the idea of the animal origin of man.

So far, this has not happened, and the discussion in Catholic circles seems not to concern the issue of human origins at all, but rather how to reconcile the plurality of first humans (polygenism) with the Catholic concept of original sin. Where does this shift in emphasis come from?

Before I answer this question, I will first point out how far Catholic scholars have departed from the encyclical *Humani Generis*. The Pope said that he did not forbid the examination of the evolutionary hypothesis, but he also added several reservations, such as that the arguments of both sides should be seriously considered. However, if we follow the current debates on evolution sponsored by Catholic institutions, we do not find representatives of "both sides." Only theistic evolutionists are invited to speak, and sometimes also atheistic evolutionists as an opposing view, because the evolutionary hypothesis is taken for granted. Meanwhile, the second reservation made by Pius XII states that the evolutionary hypothesis cannot be presented as already proven.

In addition, Pius XII writes that it is not allowed to claim that the teaching of the Church contains nothing contrary to the evolutionary hypothesis. Meanwhile, the general narrative in contemporary theological circles is that the theory of evolution does not contradict any rulings of the Church. In the encyclical *Humani Generis*, right below the sentence that makes room for studying the hypothesis of evolution, there is another one that explicitly prohibits discussion of polygenism. Think how many conferences have been organized since then, how many books written, how many articles published in Catholic journals concerning precisely the issue the Pope forbade faithful Catholics to discuss! So, today we are dealing with a huge discrepancy between the claims of theologians and the position of the Church as it has been expressed in the doctrinal documents.

Going back to the question of original sin, the reason why the discussion concerning human origins shifted from evolution to polygenism stems from the fact that if we accept the evolutionary origin of man, we must reject the historicity of Adam and Eve. This results from certain biological premises that say that the origin of the first humans could only have occurred in a population of hundreds or thousands of individuals.

And what about the Mitochondrial Eve hypothesis? Does it not prove the origin of all humans from a single pair?

For the details of this genetic concept I refer you to scientists. In 2019, an American biochemist Joshua Swamidass published a book, *The Genealogical Adam and Eve*,[7] which is probably the best account of this issue. I think it explains two things: First, why the hypothesis of "Mitochondrial Eve" does not prove the existence of a single pair at the beginning of humanity. Second, why we should not take this research too seriously. In the 1980s, when the study was first proposed claiming that everyone could have descended from a single woman living in Africa 150,000 years ago, this hypothesis somehow seemed to support the belief in Adam and Eve. But the logic of the study does not prove a single woman being the sole origin of all humans. It only estimates when all human genealogical lineages would reach the first woman who appears in the genealogy of all humans. In other words, if you go up the genealogical tree, the number of your ancestors increases exponentially—you have two parents, four grandparents, eight great-grandparents, and so forth. Since this number grows rapidly, you would quickly hit a person who appears in the genealogy of both you and your friends, neighbors, etc. The further back you go up your own genealogical tree, the more living people there are who are related to you through a particular ancestor you all have in common. "Mitochondrial Eve" would be the first woman who appears in the genealogy of all humans currently living. She would not be the "first woman" in an absolute sense, since many other lineages, many of which predated her, will have died out. She is just the first woman who happens to be an ancestor to everyone alive today.

Moreover, as time passes and different human lineages die out, "Mitochondrial Eve" and "Y-Chromosomal Adam" would also move down along the human genealogical tree. Mind that Mitochondrial Eve and Y-Chromosomal Adam are not a couple; they may appear generations apart. In fact, the Mitochondrial Eve hypothesis does not even require a single couple, just a single woman. Thus, the genetic idea of Mitochondrial Eve really has little to do with theological debates over monogenism vs. polygenism.

Is there any genetic evidence for either monogenism or polygenism?

Later, beginning with the 1990s, different genetic studies began to be used to support the thesis that humans could not have descended from just one pair. According to scientists such as Francisco Ayala,[8] the genetic diversity of humanity is too great to be traced back to a single parent in a given time. Ayala concluded that the historical Adam and Eve have been ruled out by science. It has since been shown that Ayala's research was based on biased assumptions,[9] but the myth that humans cannot be traced back to a single pair at the beginning of humanity has been perpetuated in the minds of scientists and theologians alike. It is yet another example of how contemporary theologians have been fooled by naturalists and atheists who use the alleged "scientific evidence" for their nefarious goals of contradicting Christianity.

Why should polygenism be a problem for Christians?

Because original sin is passed on from parents to children through procreation. If there was not a single pair at the beginning, then sin would not have affected all humans. This is contrary to Revelation and the teaching of the Church.

If this is so, the search for a new concept of original sin, one that would be consistent with the plurality of first parents, seems well justified.

Quite the opposite. The entire discussion concerning polygenism, apart from being "illicit" in the Church, is a theological dead end. In a few decades, we will be ashamed that theologians could ponder such things.

Why?

Because if we accept polygenism, it means that we *a priori* accept a naturalistic origin of the human body, reject the historicity of Adam and Eve, and postulate bestiality in the beginning of humanity. All of these are shameful errors. Theologians who think within these lines, should first demonstrate the evolution of man from animals, instead of simply assuming it.

On the other hand, if we accept the direct formation of the first human body by the power of God from the dust of the earth, then there is no issue with the historical existence of Adam and Eve and the problem of polygenism is null. Currently, we are struggling in the Church with errors regarding the understanding of the human soul, questioning the doctrine on original sin, and denying the existence of the first human couple precisely because the evolutionary origin of Adam's body has been universally adopted. I think that Catholics will never overcome these errors unless we return to the classic view that the description of the creation of man in Genesis is a true, historical account.

As I understand it, returning to the belief in the creation of the first man would solve many problems. But will it not also cause insurmountable difficulties? For example, it leads to the conclusion that the beginning of humanity has to be tainted with incest.

I think that this difficulty, even if it were a real difficulty, would not justify the abandonment of the faith. But, first, let's ask: What is the alternative? Evolutionists propose bestiality, because the first humans would mix with non-humans in the alleged original population. So, the only real alternative to putative incest is a bestial origin of man. Therefore, I do not think that theistic evolutionists are entitled to question the original Christian doctrine because of supposed incest.

Nevertheless, the difficulty remains, regardless of what theistic evolutionists say.

God is the Author of both moral laws—natural and divine. As He is the Author of both, He can also suspend both or make exceptions, because He is the ultimate Arbiter. This is how the Church has always understood it. For example, Aquinas teaches that Abraham would have sinned had he declined to offer his son, because this was what God directly told him to do. Surely, Aquinas adds, these kind of divine interventions, which suspend natural law, were possible only before Christ, because the perfect law had not yet been revealed to people. After Christ, no one should expect to hear from God anything against the law that Jesus revealed.

But we see an analogous situation in Eden. God tells the first human couple to be fertile and multiply and by giving this commandment, God himself suspends any law contrary to it. But I am not even sure if we can speak of actual incest in the first human family. Consider what incest is: It is a union of closely related people. However, the concept of close kinship is relative, depending on the size of the population. If one family is also the whole of humanity, then the closest kinship is also the most distant kinship. So, I think that speaking about incest in the first human family does not really mean the same thing that it means in our times.

Assuming that there were only horizontal relations (no vertical ones), the problem would appear just in the very first generation of humans. Even in our times, with these rules clearly established, you can marry your first cousin with a dispensation from the Church. Thus, the so-called incest in the first human family is really less of a problem than it is typically claimed to be by theistic evolutionists. Since there are just two options—incest or a bestial origin for man—you should go with incest anyway.

And what about the biological difficulty? Could humanity have started from a single pair? Would it not end up in deleterious inbreeding and rather quick extinction?

From a biological point of view, the first people were perfect, because they were formed directly by God. Their bodies had no defects. It can be said that Adam was a man *par excellence*, the one who most fully realized the idea of humanity in his corporeal constitution. Similarly, Eve embodied the idea of femininity to the highest degree. Their genomes contained no errors, their bodies possessed all the properties which God intended for man and woman. They could therefore give a corporeal beginning to all human races, because they contained them all in themselves. The inbreeding problem appears when there are small steady or declining populations. But rapidly expanding populations do not face this problem. Given that our first parents lived for more than nine hundred years, they could have produced more than a hundred offspring and the same would be true about their children, their grandchildren, and so on. Thus, the original human population expanded rapidly and produced genetic diversity rapidly. Humanity reached large numbers even within the lifespan of Adam and Eve. This also resolves the objections regarding Cain's wife, Cain's being the founder of a city, and other such difficulties often brought up against the original Christian belief in the single-couple origin of Man.

4. Catholicism and Evolution

When I read Catholic authors expounding on the topic of evolution, I get the impression that the Church has always supported evolutionism—even before Darwin presented his theory. What is the actual state of Church teaching on this topic?

This attitude of some Catholic authors stems partly from ignorance and partly from a strategy of wishful thinking. These theologians believe in evolution (understood as biological macroevolution) so strongly that they think they do the Church a favor by convincing everyone that it never had anything against Darwin. It seems that this attitude, at least in Poland, stems more from "bad knowledge" than from "bad will." In the USA I observe a slightly different situation. Many theologians seem to be more aware of the—let's say—lack of harmony between Catholicism and Darwinism. They understand Darwinism better, because being immersed in the English-speaking culture they have more studies and publications on the subject, and they also have a better grasp of traditional Catholic theology. Unfortunately, they support theistic evolution for various other reasons, often purely pragmatic—such as maintaining academic positions, grants, or prestige. One could therefore say that in the USA there is less bad knowledge, but somewhat more bad will.

How does supporting Darwinism relate to the teachings of the Church?

Currently, there is a significant discrepancy in the Church between what higher-ranking doctrinal documents say and what most theologians believe. That's why we need to clearly distinguish between the two. As Catholics, we are bound by the teachings of the Church's Magisterium rather than the beliefs of individual professors of theology at Catholic universities, even if some of them are bishops. It's worth remembering that bishops enjoy infallibility in matters of faith and morality only when they teach something unanimously and in communion with the Pope. In my book *Catholicism and Evolution*,[1] I demonstrated that bishops do not share one view on evolution. Therefore, the condition necessary for the infallibility of individual episcopal pronouncements is not met in this case.

Can we then say that the Church as a whole does not have a common position on evolution?

The answer depends, first, on what we mean by "evolution" and second on what we mean by "a common position." If by "common position" we mean some Magisterial proclamation, for example in the form of a conciliar constitution or clearly formulated dogma, then the lack of such a declaration is a fact. But it is also a fact that bishops, as well as theologians, differ regarding what stance the Church should take. What's more, many of them claim that the Church should not take any stance on evolution at all, because it concerns scientific issues which exceed the competence of the Church's Magisterium. A popular opinion among Catholic scholars goes more or less like this: "Let's leave scientific disputes to scientists and not confuse theology with biology."

Perhaps that last opinion is the most reasonable. Why should the Church risk a second "Galileo affair"?

As I said before, there is a fundamental difference between the case of Galileo and the case of Darwin. Paradoxically, it is the same theologians who say that disputes over evolution should be left to scientists who fail to understand this difference. Thus, we arrive at an odd situation where some theologians do not even understand which issues lie within the scope of theology and which within the scope of the sciences. Such theologians hand over the question of human origins to natural sciences —which by their nature are unable to solve this mystery. And perhaps this wouldn't be much of a problem except for the fact that theology, philosophy, the Christian understanding of creation, and ultimately the entire Christian worldview suffer as a result of this misunderstanding.

Had the early Church theologians followed such advice, the Christian faith would have been destroyed at its very origin. According to this logic, determining whether a man can rise from the dead is a matter of medicine, not the Bible or a Council of bishops under the Pope. And yet the Church defends the belief in supernatural phenomena that—though occurring in the natural order—nevertheless surpass the natural order owing to special divine power. We know that such events happened, not from scientists, but from supernatural revelation contained in Scripture. No natural science can explain them because such events lie beyond their competence.

Note that if a theologian says, "Let's leave the question of evolution to biologists," he implicitly asserts two things: First, he assumes a very questionable thesis that biological macroevolution is a valid scientific theory like, say, Einstein's general theory of relativity. Second, he implies that species were formed in a purely natural way, without any supernatural activity on the part of God. Thus, the theologian *a priori* adopts a naturalistic position that is not grounded in either his biological or his theological knowledge. It is simply an assumption stemming from his previously adopted blind faith in evolution.

Let's return to the question of Church teaching. Are there any documents condemning evolution?

Darwin's theory was explicitly condemned in a document by the Congregation of the Index issued in 1878. The decree concerned a book by Fr. Raffaello Caverni, in which he attempted to reconcile evolution with Christianity. It was a private document, meaning it was addressed to an individual, not to the entire Church. Moreover, it remained essentially hidden until 1998, when the archives of the Holy Office, which included documents from the Congregation of the Index (abolished in 1917), were opened.[2] Therefore, it is not a document of any great rank, but it is nevertheless of considerable importance when we ask about the Church's attitude towards the evolutionary origin of species.

Thanks to this document and several other similar decrees issued by the Congregation of the Index in the late nineteenth century, we know that the Church took a consistently negative stance toward Darwin's theory and toward the attempts to "baptize" it in the form of theistic evolution. It is important to notice that Caverni did not promote the animal origin of man but only the evolutionary origin of species. He even proposed polyphyletic evolution, claiming that God created several original organisms from which everything evolved except the human body. The evolutionary emergence of the human psyche was out of the question. Nevertheless, even such a limited form of evolution was rejected by the Congregation because it contradicted the separate creation of species "according to their kinds," as taught by Scripture and traditional Catholic philosophy.

Similarly negative positions were taken by the Congregation of the Index regarding two other publications—one by a French Dominican presenting views similar to Caverni's, and another by American priest John A. Zahm, who went a step further by accepting the evolutionary origin of the human body.

Still, these few private documents don't amount to much. Can one really say on this basis that the Church did not support evolution?

In response I could reverse the question and ask how many documents—even similarly low ranking—were ever published in support of the alternative position, i.e., theistic evolution? Mind that the documents issued by the Congregation of the Index were doctrinal in nature and directly addressed the question of biological macroevolution. This is something very different from modern, fairly vague statements—such as John Paul II's remark that "evolution is more than a hypothesis." Here the Pope does not even intend to make any doctrinal judgment. This phrase does not answer any question. It has a descriptive rather than a normative character. The Pope simply evaluates the current status of evolution in science, which—in his opinion—is a theory rather than a hypothesis. But he does not say whether this theory is true or false. In reality, however, evolutionists treat evolution as something more than just a theory. For instance, Michael Ruse claimed that evolution functioned as something like a religion for evolutionists, and Richard Dawkins says that thanks to Darwin's theory, it is possible to be an intellectually fulfilled atheist. So, John Paul II's statement that evolution is "more than a hypothesis" can also mean that evolution became a religion of atheists, which is, indeed, much more than just a particular scientific hypothesis.

Are there any other documents that justify the thesis that the Church did not support evolution?

There are. Mind, however, that a thesis can be rejected in two ways: Either by explicitly condemning it—as was the case with the condemnation of theistic evolution by the Congregation of the Index at the end of the nineteenth century—or by presenting an alternative positive teaching that excludes evolution not explicitly but *a fortiori*. In this way, all ecclesiastical documents that speak of the creation of the human body or the importance of the literal and historical reading of Genesis exclude evolution. Keep in mind, as well, that the Church's certainty regarding the origin of man has always been much greater than the certainty regarding the origin of species which was typically left as a matter of healthy philosophy (*sana philosophia*).

Does it mean the Magisterium of the Church has not taken a position on the origin of species?

The Church's Magisterium generally did not deal with the teaching on the origin of various forms of animal and plant life as this was considered a less important issue. The Fathers of the Church and medieval theologians maintained a common opinion about the origin of species understood as the origin of new natures (natural species). The consensus was that neither the lower natural forces nor the higher ones such as the heavens or angels could produce entirely new forms of life. Only God could derive these forms from previously created matter and He did it by His direct power.

Each species is a new concept of how the idea of life may be realized, i.e., it fulfills a plan that exists eternally in the Divine intellect. But the Church in her magisterial teachings was focused on the origin of man—not just his soul (as some contemporary theologians mistakenly believe) but also his body. Interestingly, the first centuries of Christianity were marked by debates about the origin of the soul, but the creation of the body from the dust of the earth went unquestioned until Darwin. There were, of course, some sectarian beliefs coming from Manicheans and Gnostics who claimed that the body of the first man was not formed by God but by fallen angels. But their views can hardly be considered Christian. Thus, within the Church herself belief in the creation of the body prevailed until the second half of the nineteenth century.

This belief was first defined in the sixth century by Pope Pelagius I who in his solemn profession of faith (c. 557 AD) declared:

> I confess that all men from Adam onward who have been born and have died up to the end of the world will rise again and stand before the judgment-seat of Christ, together with Adam himself

and his wife, who were not born of other parents but were created —one from the earth and the other from the side of the man.[3]

A papal solemn confession of faith is quite a document. Does it play any role in the modern debate?

I believe this one document is enough to end the debate about the animal origin of man within the Church. *Roma locuta, causa finita* [Rome has spoken, the case is closed]. However, modern theology seems to suffer from a particularly strong amnesia regarding documents that do not support evolution. When the Pope says that Adam and Eve were not born of other parents and contrasts generation with creation, it is clear that he does not speak metaphorically. "Creation" here means direct formation from the dust of the earth. This obviously excludes any kind of "hominization" of an animal.

How do Catholic evolutionists respond to such statements?

If they are aware of them at all, they usually say that these statements should not be taken literally, or that they are outdated due to the progress of science. The Council of Vienne (1312) used the Patristic expression that the Church was formed from the water and blood flowing from Christ's side, just as Eve was formed from the side of Adam who was locked in sleep. This analogy of faith (*analogia fidei*) only makes sense if Eve was actually formed from Adam's side.

One popular theistic evolutionist view is that Eve was a monozygotic twin of Adam, who developed in the womb of a hominin. But we do not say that one twin comes from the other—so I don't see how this idea harmonizes with the analogy embraced by the Council. The Church Fathers used other analogies. For example, St. Irenaeus said that human bodies will be rebuilt from the dust at the resurrection just as Adam's body was originally formed from the dust in his creation. This analogy only works if God indeed formed the first human body from the earth.

If that is a metaphor, then the future resurrection of the body is just a metaphor, as well, without any physical meaning.

The same Irenaeus and others said that Jesus was formed in Mary's womb by the power of the Holy Spirit, without a human father, just as Adam was formed from dust without a human father. Again, if the creation of Adam is metaphorical, then so is the Virginal Birth of Christ. I think that the very fact that other truths of faith were explained by reference to the creation of man from dust testifies to the unambiguous Christian position on this issue.

This belief was so strong that it served as a foundation and explanation for other Christian tenets of faith. Today, we might have to reverse this reasoning and argue for the creation of man based on faith in the resurrection and the Virginal Birth of Jesus. These statements prove that the formation of Adam from the dust of the earth was not understood metaphorically but as a real historical event.

But doesn't the claim that these statements are outdated due to scientific progress invalidate them? If biology has proven something different from what the Church taught for centuries, mustn't we admit error or reinterpret our teaching? Christianity cannot contradict science.

As you may see, this thinking is rooted in the mistaken belief that biology has proven the animal origin of man. If that were really so, I would agree that older ecclesiastical pronouncements would need some reinterpretation. But one cannot say that the Church was wrong teaching about faith since we believe the Church is infallible in matters of faith and morality. Thus, in order to believe in evolution, one needs to first reduce the origin of humanity to biology. Then, theologians who believe in "evidence for evolution" would need to say that what the Church once taught about man's origin is still important, but today we understand it more deeply, less physically, more spiritually, within the broader context of God's plan and love. Unfortunately, this kind of "escape into depth"—a kind of theological poeticism—does not solve the problem; it is more of a psychological ploy, an attempt to replace rational inquiry with irrational emotions. As for those who claim that the old pronouncements of the Church—like Pope Pelagius I's solemn profession of faith—were issued in a pre-scientific era and hence are outdated, I would point to later statements that appeared after Darwin, in the positivistic and naturalistic context of our times.

Are there any such teachings?

First there was the ruling of the Synod of Cologne in 1860, just one year after the publication of Darwin's *On the Origin of Species*. Darwin did not discuss human origins in that book, but the idea was already circulating due to other writers in Germany, France, and Italy. In this context, the local synod responded:

> Our first parents were created immediately by God. Therefore, we declare that the opinion of those who do not fear to assert that this human being, man as regards his body, emerged finally from the spontaneous continuous change of imperfect nature to the more perfect, is clearly opposed to Sacred Scripture and to the Faith.[4]

The synod first established the positive doctrine (the direct creation) and then described the opposing view. The synodal fathers defined evolution, not in biological, but in philosophical terms. Thus, it does not matter which mechanism would be proposed for evolution—whether Darwinian, Lamarckian, or neo-Darwinian—each of them presupposes gradual change from one nature—a less perfect one, because incapable of receiving the rational soul—to a more perfect one, because capable of receiving the rational soul.

The synodal formula excludes such a scenario. The decrees were first approved by the Roman congregations and then separately recognized by Pope Pius IX. So, although it was a voice of the local Church, we are assured that it was in full harmony with the attitude of the Apostolic See and the Pope himself.

Even more significant is a section of Pope Leo XIII's 1880 encyclical *Arcanum divinae sapientiae* [The Hidden Design of the Divine Wisdom]. The Pope refers to the opponents of the creation of Adam from the dust of the earth and Eve from Adam's side as the "enemies of the faith," who are unwilling to accept the "unchangeable Church teaching." The Pope confirms that Adam and Eve constitute the natural beginning of the entire human race; thus he excludes polygenism.

Another important doctrinal document is the 1909 "Response" of the Pontifical Biblical Commission regarding the historicity of the first three chapters of Genesis. Among several truths that must be regarded as historical facts, the Commission lists: "the special creation of man" and "the formation of the first woman from the first man." "Special creation" is clearly something different from "natural evolution." The decrees of the Biblical Commission exclude evolutionary scenarios by positively endorsing the alternative view which is special creation.

Does reading the text of Genesis non-literally necessarily amount to supporting evolution?

As we said before, Genesis contains several parallel senses, but it also contains many elements that are literally and historically true. In the 1909 "Response," the Biblical Commission ruled that in the Genesis account of creation there are truths that need to be taken as history. The Commission listed five such truths, among them the special creation of man. Surely, even in these passages one could seek mystical or allegorical meanings, but that does not excuse anyone from believing in the historicity of these events. In other words, there was such a place and time when Adam appeared on earth by the working of Divine power. Surely, the Tradition of the Church discovered a moral meaning, as well in the formation of Adam from the dust of the earth and Eve from Adam's side. Thus, we can see that a non-literal reading of the Biblical text is not equivalent to the acceptance of evolution.

Let's return to the Church Fathers. From what you said it follows that they spoke of the formation of Adam from dust. Modern theologians, however, often invoke the Church Fathers when they defend so-called "Christian evolutionism." When St. Basil, for instance, says that the world is an unfinished masterpiece—does that imply he supports evolution?

Theistic evolutionists readily quote St. Basil, but even more often St. Augustine. They claim that because Augustine was an evolutionist, so evolution and Christianity go well together. But they forget to add that Augustine and Darwinists use the word "evolution" in entirely different senses.

In what sense did Augustine speak of evolution?

Augustine was influenced by Stoic philosophy and Manichaean errors. One of the problems with pagan concepts of the formation of the world was that the deity does not make the world perfect, but somehow improves it, tweaks it, and even then the world, men and gods remain flawed and imperfect. When Augustine became familiar with the biblical account of creation, he could not understand how God would create the world subsequently over time, that is, over the course of six days. The idea of many creative acts that somehow bring the world to perfection reminded him too much of pagan cosmogonies. And that is why he sought some metaphysical model that would allow him to interpret the biblical account in such a way as to preserve the perfection of Divine action in creation. To his aid came the Stoic concept of "seminal reasons" (Greek *logoi spermatikoi*), that is, the idea of certain seeds that God created at the beginning and which later matured and developed to achieve the fullness of their natures. Augustine therefore held that God created the whole world in the beginning in one moment, although some things only in the form of these seeds. This concept looked quite neat and fitted well with Augustine's Greek rather than Hebrew mentality.

Augustine provides a sophisticated interpretation of Genesis. But does it really harmonize with the text?

Of course, Augustine was aware that his interpretation diverges from the text. Moreover, he repeatedly asserted that he did not know how to interpret given fragments of the creation account. He solved the problem of the "six days" in a rather pragmatic way by stating that Moses divided into days what happened at one moment, because otherwise the crude recipients of this text would not understand it. So, you see how Augustine's Greek, abstract, and somewhat over-intellectualized approach displaced the more original interpretation, present both in the Hebrew tradition and in most of the Church Fathers, for example in St. Ambrose.

Eight hundred years later, St. Thomas Aquinas while synthesizing Christian theology would say that we have two traditions of interpreting Genesis in the Church. One descends from Ambrose, the other from Augustine. Ambrose's tradition advocates the gradual shaping of the world by God over a period of time, which Holy Scriptures call six days. In turn, Augustine's tradition speaks of the creation of everything at one moment, although some things only as seeds, which later, after completion of creation, are to develop to the fullness of their natures thanks to the powers inherent in nature. Thomas also says that the tradition of Ambrose is maintained by most saints and Fathers and is more in line with the Bible; unfortunately, however, it is also more susceptible to attack by unbelievers. Augustine's tradition, on the other hand, is shared by a minority, harmonizes less with Holy Scripture (at least as first glance), but at the same time is more resistant to criticism from unbelievers. Interestingly, Thomas declares that he would defend both traditions, and in another place he states that these two traditions differ only in non-essential points.

So it seems that there is after all a place for evolutionism in Christianity. Although this tradition initially was shared by a minority, it has been present since the beginning of advanced theological reflection.

Yes, but in the Augustinian tradition evolution means something different from what it means in Darwinism. First of all, Augustine maintained that God created all things in the beginning according to their kinds. Therefore, even if animals and plants were created in the form of seminal reasons, these "seeds"—however we are to understand them—already contained within themselves the nature or essence of a given species. In Augustine there is neither the idea of common descent from a single ancestor, nor natural transformation of species, i.e., the transformation of one kind of living being into another. Each individual nature comes directly from God.

Augustine also excluded the possibility of any entirely new nature being formed by a created entity, be it material or angelic. Therefore, the word "evolution," which he used when speaking of the growth of the "seminal reasons," means no more than microevolution, i.e., changes within species. Nothing fundamentally new can arise through evolution understood in this way. This is *e-volution* (from Latin *e-volvere*) in the strict sense, i.e., the development of what already exists in an *in-volved* (hidden) form. Darwin himself was aware of the fact that the word "evolution" did not mean what he had in mind. And that is why he did not use the word in his early works. It was Herbert Spencer who expanded the term "evolution" to mean "creative evolution," that is, the process that leads to the emergence of various novelties. And only when this new understanding of the word had settled in the popular perception did Darwin introduce it into his works.

Is the Augustinian interpretation of the creation account still relevant?

Today we know that different species were appearing fully developed over long periods of time. Therefore, they did not arise all at once at the beginning, as Augustine thought. Neither do we find anything in the fossil record that could stand for the "seminal reasons" that he imagined. In the context of our current knowledge about the natural history of our planet Augustine's interpretation is untenable.

However, two qualifications must be added to soften the tone of this statement: First, phenomena such as the Cambrian Explosion show that there were moments when many species appeared suddenly and simultaneously (in a geological moment).

Second, microevolution, i.e., environmental adaptations of various organisms, is a fact. Prof. Lenartowicz spoke of the "totipotentiality" of living organisms, i.e., their innate ability to respond unconventionally to environmental influences. These responses happen by virtue of the laws of life without any supernatural action of God or angels. They therefore fulfill the condition of development within the lines of the "seminal reasons" that Augustine spoke of. Thus, the Augustinian interpretation of Genesis may still be valid in some limited sense.

From what you say it follows that the separate creation of species was the position of all Christianity, including Augustine, even if individual authors differed in their interpretation of the Book of Genesis. But let's get back to the present. Hasn't the position of the Church evolved on the issue of the origin of life and man?

Certainly, the mid-twentieth century saw a change in the approach of Catholics to evolution. This is most evident when we look into the catechisms. The first catechism, the so-called Tridentine (or St. Pius V's) catechism, and all local catechisms published before 1950, taught explicitly that the first human body was formed from the dust of the earth. In the Catechism of St. Pius V, when it talks about creation, we read:

Lastly, God formed man from the dust of the earth, so created and constituted in body. ... While Adam was asleep, [God] took one of his ribs and filled it up with flesh. Then the Lord God formed the woman from the rib which he had taken from Adam.

Another example is the Baltimore catechism of 1885, which explains:

On the sixth day God created man and called him Adam. ... God could have made Eve as He made Adam, by forming her body out of the clay of the earth and breathing into it a soul, but He made Eve out of Adam's rib to show that they were to be husband and wife.

We see, therefore, that a merely symbolic understanding of the biblical text is not an option; the catechisms spoke about the actual manner in which the first people came into being.

The so-called "new catechism" of 1992 is the first to stay silent on the question of how humanity began to exist. In no. 362, we read:

The human person, created in the image of God, is a being at once corporeal and spiritual. The biblical account expresses this reality in symbolic language when it affirms that *then the Lord God formed man of dust from the ground, and breathed into his nostrils the breath of life; and man became a living being* (Gen 2:7). Man, whole and entire, is therefore willed by God.

Of course, such a formulation is "grist for the mill" for theistic evolutionists. They believe that the Church confirms that the biblical account of the creation of man should be understood symbolically. The authors of the catechism suggest that Gen 2:7 conveys only a very general truth of a moral and spiritual character, namely, that the whole man is willed by God. This is an obvious reduction of the metaphysical content to the moral content alone.

Can catechisms contain errors?

The catechism by itself is not the source of the Church's faith; it is only supposed to refer to it, i.e., to say what the Church believes. The source of faith is the Holy Scripture, conciliar constitutions, papal pronouncements, etc. It may therefore happen that a catechism does not correctly convey what the Church teaches. It is then said that the catechism contains an error, even though the Church's teaching remains infallible.

From the examples you cite, does it not follow that the Church has changed her position on the issue of the origin of the human body?

Catholics who believe that the Church has accepted evolution rarely notice important distinctions that must be taken into account. First, do recent statements by popes and various theological bodies abrogate earlier teachings? Second, what kind of evolution has the Church supported and how far does the support go? Finally, we need to ask whether the change of the attitude in the Church is justified and whether it is a permanent or merely temporary state of affairs? The quoted passage from the new catechism does not say that man came about through evolution. It only implicitly rejects the historical and literal interpretation of Genesis 2:7. If the catechism were to state directly that this verse must not be interpreted literally, it would contradict the teaching of the Pontifical Biblical Commission of 1909, which acted in the name of the Pope (the Magisterium of the Church) and whose decisions have never been revoked. Therefore, in my opinion, the new catechism does not fully and faithfully report the teaching of the Church on this particular issue.

Do you think we can ignore the support that John Paul II gave to evolutionism?

First of all, John Paul II did not support the idea of the simian origin of man at all. In my opinion, John Paul II did not present any specific teaching on the evolutionary origin of species, either. As we know, his texts cover thousands of pages and contain many in-depth analyses in various theological fields. But all of his statements on evolution can fit on one page. This is very little and so it is difficult to reconstruct with full certainty John Paul II's attitude to evolution. In one of his Wednesday catecheses delivered in 1986, the Pope said that "the theory of the evolution of nature, understood in a sense that does not exclude divine causality, is not in principle opposed to the truth about the creation of the visible world, as presented in the Book of Genesis." I think this is the most "pro-evolutionary" statement by John Paul II. Nonetheless, it did not resonate in any significant way in the Catholic media. It is also rarely cited by theistic evolutionists.

Why do you think that is the case?

The only reason I can think of is that this statement assumes the normative nature of Holy Scripture. So, although John Paul II said that the theory of evolution of nature does not contradict the biblical account of creation, it follows from this statement that theories created by scientists could contradict Holy Scripture and thus they could be dismissed based on the Bible. This implies quite a classical understanding of the Bible as an authoritative text in explaining the questions of origins, which is not what theistic evolutionists want to admit.

In addition, in this statement it is not clear what the Pope means by the "theory of the evolution of nature," nor is it certain what the words "in principle" mean. If we understand evolution as simply change over time, then surely Holy Scripture does not contradict it. We observe changes in nature every day, and I think that "the evolution of nature" in this sense is not controversial.

The question that is controversial is whether the changes we observe can explain absolutely everything, and especially the origin of species. In addition, John Paul II added another condition in this statement, namely, that the theory of the evolution of nature cannot exclude divine causality. And this is the core of the dispute between evolutionists and creationists. Theistic evolutionists do not exclude divine causality in some general sense, but they exclude direct causality in the creation of a new species. So, if the Pope had in mind this type of causality, then theistic evolution does not meet the condition of acceptability presented by John Paul II. And Darwinism, which excludes any divine causality, certainly does not meet this condition.

Can we say that John Paul II accepted evolution in the sense of, for example, common descent, but that we do not know what he thought about Darwinism as a "mechanism" of evolution?

Darwinism says that variations that determine the emergence of biological novelties, including humans, are entirely random, while their "creative capacity" is determined by necessity in the form of natural selection. The more prominent Darwinists constantly emphasize the random and purposeless nature of the evolutionary process.

Meanwhile, in a catechesis delivered in 1985, John Paul II said:

> To all these 'indications' of the existence of God the Creator, some oppose the power of chance or of the proper mechanisms of matter. To speak of chance for a universe which presents such a complex organization in its elements, and such a marvelous finality in its life would be equivalent to giving up the search for an explanation of the world as it appears to us. In fact, this would be equivalent to admitting effects without a cause. It would be to abdicate human intelligence which would thus refuse to think, and to seek a solution for its problems.

Yet Darwinism opposes the work of creation with the internal mechanisms of matter.

It's hard to tell what else other than Darwinism the Pope might have had in mind here. We can therefore only speculate about what John Paul II thought about biological macroevolution; but we should conclude that he rejected Darwinism as understood by modern science.

Are there any boundaries that the current teaching of the Church respects regarding the acceptability of evolutionism?

I think that these boundaries are outlined in the quoted passage from John Paul II. Very similar ideas can be found in the teachings of Benedict XVI. They contain certain concessions for "guided evolution," but they certainly do not allow for unguided, that is, Darwinian, evolution. Recent popes have unequivocally opposed the "hegemony of chance."

Note, however, that in a way this is a paradox. Until today, the only well-defined mechanism of evolution, as biology understands it, is the Darwinian mechanism or one of its variants. However, all of these mechanisms have in common that they are blind and unguided. Therefore, recent popes have given implicit consent to evolution, but under conditions which no currently proposed macroevolutionary theory meets. Only the theory of intelligent design (ID)—which allows for species transformism but excludes the possibility of chance's being the main cause of all biological diversity—could be accepted in light of what the recent popes have said.

Do the past fifty years constitute a revolution in the Church's teaching on evolution?

No, because the Church has not made any significant rulings on this issue in the last seventy years. All statements ranked very low in the hierarchy of ecclesiastical documents, and none of them were intended to present a definitive teaching of the Church. There are also some documents produced by bodies such as The International Theological Commission and The Pontifical Biblical Commission (after 1971) which are more or less supportive of evolution, but these are not Magisterial documents; they do not belong to the teachings of the Church and should not be quoted as such.

Finally, I wanted to ask about contemporary theology. You spoke earlier about the discrepancy between the Church's teaching on evolution and the position of theologians. Do you see any positive contribution from authors such as Pierre Teilhard de Chardin? He believed that the entire world is encompassed by a great cosmic evolution that is heading towards divinization. God has already

revealed himself to man, and now the world is evolving to the
Omega Point—the Parousia, the return to God.

I think that the theology of Teilhard de Chardin is already a thing of the past. It is true that his writings greatly influenced what we could call "post-conciliar theology." But this fashion, or rather "seduction," passed away with the generation of '68.

That generation was in love with a type of theology in which many traditional Christian concepts acquired a new understanding that was strange to the Catholic tradition. I think that Teilhard was a master at redefining old concepts and creating new ones. But it does not seem that he is still shaping Catholic theology. In vibrant theological centers, such as those we find in the United States, it is more common to study authors such as St. Thomas Aquinas, the Fathers of the Church, and the classics of contemporary theology, such as Hans Urs von Balthasar or St. Pope John Paul II.

As to Teilhard's legacy, it must be said that the Church opposed his method and research many times and in many ways during his lifetime. But these were mainly private prohibitions and admonitions. The books of the French Jesuit became famous only after his death, when—contrary to the decisions of the ecclesiastical authorities—they began to circulate in departments of theology and seminaries. Some people were delighted with these works, while others showed extreme reluctance.

At some point, the attitude towards Teilhard certainly became more than an attitude to just one of many scientist-theologians. It became in a way a litmus test of the entire worldview that a given person represented. If you liked Teilhard, then you were scientific, modern, progressive, and open to culture and the experience of transcendence. You did not like him? That meant you were an obscurantist, a conservative, a Thomist.

I think that everyone took Teilhard too seriously. His theology is not theology in the strict sense, just as his science is not a serious science. I think that Étienne Gilson very aptly described Teilhard's works as "theology-fiction." The 1970s were a time when literature called science-fiction gained popularity in mass culture. This literature referred to concepts of the real world, but it was a fiction, something imagined. Teilhard became popular because he filled a market niche for a theology-fiction type of literature. The concepts he created, the ideas he developed, referred to the real world and Christianity, but in fact they were a projection of his imagination; they did not describe the history of creation, nor the history of salvation, nor even eschatology. I think that Teilhard's works should be put on the shelf between Tolkien and Stanisław Lem. You can admire them, but you should not take them seriously.

5. Are Creationists Simply Ignorant?

For many people, both scientists and theologians, creationist views are an object of ridicule. Why is this?

The reason for this state of affairs is the unconditional victory, not only of the Darwinian theory of evolution in science, but, more broadly, of the evolutionary paradigm in contemporary culture. If you are born into a culture where you are taught evolution from childhood and you later meet someone who does not share the same beliefs, his views may surprise you.

Contempt, however, is something else. If contempt appears, it is a problem of wicked will. I will point out, however, that creationist views, as far as I can tell, are an object of laughter or contempt mostly in Europe. I do not see that much of it in the USA. Of course, this does not mean that there are no opponents of creationism in America. It is simply that creationist views have been propagated mainly in the USA, so they do not meet with the same level of surprise and incredulity as in Europe.

I suppose that many educated Americans meet over the course of their lives other educated Americans who believe in Young Earth Creationism. If you meet such a person and see that he/she is a decent citizen, and oftentimes a strongly dedicated believer, you naturally become familiar with their ideas. This does not mean that you embrace them. You simply become more tolerant.

So, can creationism be taken seriously?

Here, we must make a necessary distinctions. As St. Thomas said: *Distinguere sapienti est* [Making distinctions belongs to the wise]. I think that the claim that the Earth is less than ten thousand years old cannot be taken seriously. The scientific data that point to a much older Earth is simply overwhelming. Young Earth Creationists often argue that the methods of age testing are inaccurate and indicate discrepancies in the results. The problem, however, is that we are talking about differences in orders of magnitude rather than small inaccuracies. If all the methods point to an age of millions and billions of years, then the discrepancies do not matter because we are still very, very far from a few thousand.

Young Earth Creationists invested much effort into dismissing the scientific evidence and data for the old universe. And I think this is the reason why there is an aversion to them in scientific circles. It is difficult to deal with someone who is notoriously mocking and abusing science in order to squeeze it into his religious assumptions.

Isn't the belief in the direct and separate creation of species, outside of the laws of nature, the same "abuse of science," basically amounting to a belief in magic?

This is a different question. Let's distinguish between two issues: the age of the universe and the origin of species. The question about the age of the universe, or of the Earth, is a scientific question, while the question about the origin of species goes beyond scientific knowledge.

Let me remind you of the rule I already expounded: Check what kind of question a theory addresses and you will know whether it is scientific or not. The question of the age of the universe is not about where something came from. It is a question about the current state of affairs. It says how old a given structure is today. Therefore, it is not a question addressing the origin of a thing and so it remains within the scope of science. On the other hand, the question about the origin of species belongs to the type of questions that concern the beginnings of things; such questions ask, Where did something come from? And, therefore, science cannot provide a complete answer to them.

If the belief in the creation of species is a kind of childish belief in "magic," then all the more must we recognize as such the belief in the miracles of Jesus, such as the raising of a dead person to life and miraculous healings, not to mention the Eucharist, the future resurrection, and practically everything the Church teaches that goes beyond the order of nature. I must admit that I do not understand why the belief in the separate creation of species is so hard to accept by so many Christians who easily accept all those other, much more incredible, events. At the same time, we have no better explanation for the origin of species.

Evolutionists would not agree. They believe there is a natural explanation of the origin of species. Is the creationist theory better than the evolutionary theory?

People who ridicule the belief in the supernatural formation of various forms of life are convinced that they have a better theory—evolution. It seems to them that this theory is better because it does not require supernatural action on the part of God. In the nineteenth century, when people were still delighted with the discovery that angels do not have to hold the planets up to keep them orbiting, naturalistic explanations seemed better. But notice the kind of reasoning that pervades this mindset—the theory of evolution is better because it is naturalistic. Not because it explains anything, not because it orders any data, but because it does not need God.

Naturalists are so delighted with the naturalism of their theory that they completely miss the fact that it hardly explains anything. They prefer to have a façade of a theory instead of a real explanation simply because this façade frees them from appealing to the supernatural. When atheists take such a position, it is somewhat understandable. After all, they must believe in evolution no matter whether there is any evidence for it or not. But when the same naturalistic position is adopted by Christian theologians, who are responsible for transmitting the faith without distortion, then something is wrong.

> *But if we accept the creation of species, doesn't that spell the end of science? After all, we can always say that God did it. No one will look for natural causes of natural phenomena anymore.*

Do you really believe that we are in danger of such a scenario? I do not think so, for the reasons I mentioned a moment ago. We have a clear criterion for discerning which questions are scientific and which are not. And this criterion cannot stem from science itself. No one is a judge in his own case. Just as physics does not establish the laws of physics and biology does not explain the origin of life, so science as such cannot define the boundaries of science.

Note that 99% of scientists do not need the theory of evolution for anything. Each of them works in their own very specific field, in which they have specific questions, experiments, and problems. We can say this: A physicist studies particles, atoms, and quarks, splits particles, adds up energies, and studies radiation, but the question of where matter and energy came from in the first place does not need to bother him anymore than any other person.

If the same physicist in his free time feels like learning more about the philosophy of nature, he will pick up a popular book on cosmology and read about the formation of the world. Unfortunately, he usually comes across a completely naturalistic story somewhat reminiscent of pagan mythologies, with the only difference being that the names of deities have been supplanted with scientific terminology.

Similarly, a biologist studies proteins, genes, cells, and cell organelles. He spends entire days in front of a computer comparing sequences of nucleotide bases. This is his world. But where life or species came from in the first place is not his question. He does not need to have any greater interest in this question than anyone else. He was told already in elementary school that all of this came about through evolution. Typically, a biologist knows as much about evolution, or just a little bit more, than his colleagues working in business or the media. He believes in the theory of biological macroevolution because "this is what they say," this is what is written in the textbooks. He himself has never had the need to check whether this assumption has been proven. He trusts his colleagues working in the departments of "evolutionary biology," i.e., institutions specifically designated to perpetuate evolutionary materialism.

Biochemist Michael Behe once said that he went through his entire education, from grade school to his doctorate, and he never heard any arguments against Darwinism. The schools he attended were Catholic schools which taught the standard version of theistic evolution: "If God wanted to use evolution, He could have, but the most important thing is that God loves all of creation." It was only after he got his doctorate in biology that he came across a book by Michael Denton *Evolution: A Theory in Crisis*.[1] After reading it, he realized that there were powerful scientific arguments against Darwinism. Behe had been taught by his catechists that there were no religious problems with evolution. He is not a theologian, so he does not want to judge this issue. But he wonders why these catechists completely ignored the scientific problems of the theory they so blithely accepted.

Would you say that creationism is not a threat to science, then?

If we understand creationism solely as a belief in the supernatural actions of God that led to the formation of the fundamental elements of the universe and ended long ago with the creation of man, then not only does it not pose a threat to science, but it actually helps science to find its limits and scientists to open their minds to the supernatural.

A threat to science is Darwinian evolution, which is maintained in contemporary biology despite empirical data only because naturalists have not yet developed any other naturalistic theory that could replace the neo-Darwinian explanations and avoid the now unavoidable reference to intelligent design. The theory that was so boasted about only a few decades ago, when little was known about the informational nature of DNA, has now become an obstacle to progress in biology.

There are specific examples of this, such as the discovery of function in the so-called junk DNA, which would not have been possible had scientists consistently followed Darwinian assumptions. This and other examples show that Darwinism is an obstacle to scientific progress. I think a similar problem affects physics, where the multiverse theory is promoted only to avoid the theory of cosmic fine-tuning. Just as in biology, so in physics, naturalists are telling us to abandon science in order to avoid inconvenient conclusions that follow from science itself. And this is the real threat to science.

Are we justified in abandoning the evolutionary paradigm at this point? That would mean changing almost all the textbooks. Would this require some kind of revolution in science?

Mind that the macroevolutionary theory, whether it concerns biology, chemistry, or physics, always functions, as it were, on the margins of science. It is said to be the foundation of modern science. But evolution is discussed in separate chapters in textbooks precisely because it does not fit scientific questions. It is a certain worldview within which we are told to look at all biological phenomena. It is said that this theory comprehensively organizes scientific data, helping to order them in a coherent system. I doubt that evolutionism actually does this, but even if it did, it is not important whether a theory "looks nice" but rather whether it is true, that is, whether it corresponds with what actually happened in the past.

Evolutionists delighted with the "coherence of evolutionism" completely fail to notice that they are moving in spirals of circular reasoning: The theory of evolution is coherent because it organizes facts from different fields. And why does it organize facts? Because it is coherent. It is naturalistic and therefore scientific. And why is it scientific? Because it is naturalistic. It is the paradigm of modern science, because it is accepted in all fields. And why is it accepted in all fields? Because it is the paradigm, etc. In fact, we can trash this entire paradigm and solid science will remain solid science. Great discoveries in chemistry, biology, and physics will remain great discoveries, because they do not depend on the answer to the question of where it all came from.

The fact that the first textbook on intelligent design (*Of Pandas and People*[2]) was created by automatically changing the word "creation" to "intelligent design" was once laughed off as a joke. I encourage you to do something different. Take any biology textbook and consistently change the word "evolution" (used in the sense of biological macroevolution) to the word "Creator." You will notice that the solid science presented in the textbook remains solid science. Science simply does not need the general theory of evolution for anything.

Is there such a thing as a theory of creation?

Let's begin with explaining that the term a "theory of creation" is not quite accurate. It suggests that explaining the origin of species by referring to creation is a scientific theory that can compete with other theories, such as Darwinian evolution. In fact, however, belief in creation is not a theory at all. If it is formulated within a discipline such as theology, it is a theological concept.

Calling belief in creation a "creation theory" or "creationism" suggests that we are confusing the planes of natural science and theology. Creationists themselves often fall into such a confusion and this is why they call their belief "creationism" or even "scientific creationism."

What different types of creationist views are out there?

In the Church, or in Christianity more broadly, we currently have three views on the origin of species. All concepts can be classified into one of these categories: (1) theistic evolution, (2) progressive creation, or (3) Young Earth Creationism.

Theistic evolution is sometimes called "evolutionary creation." This name is an oxymoron akin to "scientific creationism," because if something was created it could not have evolved and if it evolved, i.e., emerged naturally due to the workings of nature, then it was not created. Of the three views mentioned, only the second and the third are creationist. Theistic evolution is the belief that God did not create species by his own direct action, but instead used natural processes which are together called "evolution." This view does not recognize God's supernatural activity in the natural world after the first act of creation by which God created space and matter out of nothing (*creatio ex nihilo*).

Theistic evolutionists usually do not rule out the possibility that God could act supernaturally, but they claim that He never did so when it comes to the formation of the universe. Theoretically theistic evolution does not need to adopt Darwinism as a mechanism of evolution but in practice it does, because Darwinism is still the mechanism most commonly accepted in biology due to the lack of any better naturalistic alternative.

The creationist concepts have in common the belief in the supernatural action of God in the formation of the universe, specifically in the origin of species. To this idea, Young Earth Creationists add the belief that the universe is no more than 10,000 years old and that it was created in six days, which they understand as natural days (six periods of 24 hours). Biblical fundamentalists adopt the young Earth perspective, but they justify it with the text of Genesis understood as the only source of knowledge about origins. Scientific creationists, on the other hand, attempt to justify Young Earth Creationism with scientific data.

How does Catholic creationism differ from Protestant creationism?

Let's begin with acknowledging the fact that every Christian is a creationist. Creationism in its basic sense is a belief in creation. If someone does not believe in creation, he/she is not a Christian. Indeed, Catholic creationism, as we find it in the Church Fathers or medieval Scholastics, differs quite significantly from Protestant creationism, which originates in the physico-theology of the seventeenth century. I will point out a few basic differences.

First, Protestant creationism understands Divine action as "interventionism," a series of catastrophes, the entry of God's omnipotence into the order of nature. In the Catholic understanding of creation, however, the emergence of new species by Divine action is not an intervention. Intervention (from Latin *inter-venio*) is God's entering into the natural order of causes and effects and producing more or less dramatic changes in it. Catholics understand the formation of the universe as the addition of beings, without destroying anything that existed.

St. Ambrose compares Divine action in creation to an artist painting an image. When a painter paints a landscape, first we have a meadow and sky. And this picture is complete in itself, it lacks nothing—it is good. But the painter can add the sun and stars or clouds and the picture will be even more beautiful. He can also add birds in the sky, animals on land, and fish in the water, and then the picture will become even more beautiful. But adding these adornments does not entail a destruction of anything that existed before. It is not the erasure of something that was before, scratching holes in the canvas.

Christian tradition called the creation of stars, planets, and plants the work of distinction (*opus distinctionis*) and the creation of animals the decoration of the world (*opus ornatus*). There is no shadow of "interventionism" here, although there is clear teaching about the direct and supernatural creation of plants and animals according to their kinds.

Does this mean that Protestants understand creation differently?

It is hard to consider all Protestants as a uniform group, but undoubtedly in most traditional Protestant creationism the act of creation itself is understood differently. In the Catholic tradition, only God can create. No creature, whether angels or material causes, can serve as God's help in creation. And this is the second difference. I think that neither Protestant fundamentalists nor Catholic evolutionists understand this point. One could endlessly quote the Fathers and Doctors of the Church saying that the act of creation can only be a direct action of God. But I think it is more important to understand why this is so: In every act of creation, something new is created that had not existed before. If this is the first act of creation described in the first verse of the Bible ("In the beginning

God created the heavens and the earth"), then space and matter come into being out of nothing. If these are subsequent acts of creation—for example, those acts by which new living beings come into being—matter already exists, but God creates a new being by introducing a new form, that is, a new shape, structure, concept, definition, or idea that did not exist in this matter before.

Therefore, in the case of the secondary acts of creation, the matter of being does not come into existence from nothing, but a new being comes into existence from nothing by the education of a new form from the previously existing matter. Therefore, in every act of creation there is the creation of some absolute novelty, something that did not exist before, and now it begins to really exist.

There is no proportion between nothing and something. And that is why only an omnipotent being by its own power can overcome this infinite distance between non-being and being, which means He can emanate something from nothing. And that is why only God can create.

Theistic evolutionists who deem evolution the "secondary cause of creation" do not know or do not understand what creation is in the Christian sense of the word. A secondary cause—anything that is not God—can only act on something that already exists. But changing something that already exists into something that has never yet existed is not creation, but transformation. This is why there are no secondary causes in creation.

Are there any other differences between Catholic and Protestant creationism?

Another difference concerns the understanding of the relationship between science and religion. For many Protestants, scientific data constitute "evidence of creation," but for Catholics not necessarily. Since creation is a direct (and therefore supernatural) act of God, it is clear that it escapes any scientific scrutiny. Science can only say that it comes to the limits of natural explanations. It can therefore provide a "negative background" by stating: "we know how the thing changed until this point; we do not know where it came from".

For example, paleontology can determine when and where which dinosaurs lived. But it cannot say where they came from. The limits of scientific explanation apply to both evolutionists and creationists. The former cannot claim that scientific data prove evolution, the latter cannot claim that scientific data prove creation. For Catholics who accept the traditional understanding of creation, scientific data are simply consistent with the Bible, but they do not by themselves prove its veracity. Faith is always faith, that is, a supernatural act of will that leads reason to accept truths for which there is no empirical evidence.

Another significant difference concerns the attitude towards the "young Earth." I already mentioned that the Catholic Church ruled at the beginning of the twentieth century that the word "day" from the account of creation could be interpreted as a period of time different from a natural day. Therefore, Catholics may, but do not have to, be Young Earth Creationists. It was a very mature ruling of the Church which stemmed from thorough reading of the theological tradition. This ruling shows that the Church did not have a problem with true scientific findings, such as evidence for the old universe.

Catholics are somewhat immune to Young Earth Creationism because they recognize Holy Tradition as the milieu in which the Bible should be interpreted. Barring Tradition, only one of two options remains for biblical interpretation: either the letter of the text or liberal exegesis. The former leads to fundamentalism, the latter to the deconstruction of Holy Scripture. Those Protestants who strive to avoid deconstruction typically end up with the concept of a young Earth.

I think that the attempt to prove creation using scientific data stems from a more general problem in Protestant theology which puts into opposition the Bible and science, faith and reason, grace and nature, human freedom and Divine will (theology of predestination), God's mercy and God's justice, etc. This attitude finds its roots in the theology of the Reformers. The Catholic approach is more synthetic. We can say that Catholicism is guided by the principle of "both-and" instead of "either-

or." And that is why Catholic creationism recognizes reliable scientific data. It also recognizes the value and place of science in our understanding of reality, but at the same time it maintains divisions between the different levels of discourse.

There are some Young Earth Creationist arguments that seem to be mainly theological, as well. For example, some proponents of the young Earth view say that allowing billions of years of natural history of the universe, necessarily means that animals died before original sin. They believe deep time and ancient extinct species contradict the passage in Romans, chapter 5, that says death first entered the world through Adam's sin. How should we understand an ancient universe in connection with original sin and death?

I think this argument contains a twofold problem. One is that death in St Paul's words refers to human death alone. Thus, animals could have died before Adam sinned, and this view is reconcilable with the old earth view. Another problem is that the short time or young Earth view of divine creation actually does not make any animal death impossible. On the contrary, the more we think about it, the clearer it becomes that without fundamentally changing all of nature such an option is not possible. Any animal while moving around would accidentally destroy some insects, any organism would have bacteria that eat flesh or parasites, any digestive process would destroy bacteria and parasites. It is hard to imagine that predators would refrain from predation even for a moment.

Moreover, if we take St. Paul's words as referring to animal as well as human death, it follows that not only did animals not die before sin, but they were immortal in the same way as our first parents were (by special divine grace). But this is never stated in the Bible and it would not make much sense in the light of the divine plan for the universe.

Some creationists proposed that the animal kingdom, when initially created, was totally different in its nature, so that there was no death. But the Bible does not say anything like that. We do not read about any transformation of all of nature after the fall. This kind of idea resemble

Gnostic visions in which the material universe is naturally evil and subservient to the evil gods. On the Christian account, nature was created good and it did not change at the moment when humans sinned.

The Bible speaks about the consequences of the first sin as applying to humans—their death, pain, labor, but most of all their falling out of grace. So, the greatest consequences of human sin are actually spiritual, not material. For these reasons, there is no argument in the Letter to the Romans for the young Earth point of view.

Creationist books are often full of scientific data and arguments. Is creationism a religious or scientific theory?

Creationism is a religious concept present primarily in Judeo-Christian theology. Resorting to scientific arguments to prove creation leads to confusion of the different levels of discourse. If the supernatural activity of God can be scientifically proven, i.e., directly observed or shown by an experiment demonstrating that a supernatural power was at work, then this power would have to be a part of nature. So, either God is a part of nature or His action cannot be observed in a scientific way.

Of course, observing *the effects* of Divine action is something else. For example, if we see that a boy has a few loaves of bread, and then we see that these loaves fed several thousand people and several baskets of leftovers were collected, then we cannot explain these phenomena based on the laws of nature. If not the laws of nature, then what kind of explanation could there be? A person open to the possibility of supernatural causation will assume that the power of God acted here in an invisible way. However, this explanation is no longer a scientific one, but a religious one. We know from the Gospel that this is what actually happened.

This does not mean that this explanation is unreliable, subjective, or meaningless. In fact, theological explanations are inherently more certain than scientific ones. But according to naturalists, meaningful explanations come only from natural science. A person closed to the possibility of Divine supernatural causation would say that currently we do not know how to explain the miracle of the multiplication of the loaves, but one day science will find an explanation.

It is similar with creation: By observing nature, we conclude that new species appear suddenly, as if out of nowhere; they exist for millions of years pretty much unchanged, and then they die out. Science has no explanation for this phenomenon. But does this mean that we observe creation in the fossil record? I do not think so. The fossil record provides data that do not exclude creation. That is all that science can deliver, and this is enough for us. We ask religion for positive explanations.

6. Is Evolution in Trouble?

Charles Darwin was not the first naturalist to address the evolutionary origin of species. How was that idea viewed before Darwin?

Indeed, the idea of a common descent of species from a common ancestor by natural generation was invented long before Darwin, in the early modern period. Descartes claimed that such an idea would make it much easier for us to explain the origin of species, although he did not push it because this idea contradicted the biblical faith. I think that Descartes's brief remark in the *Principles of Philosophy* (1644) shows well what motives governed evolutionary thinking from the beginning. It was all about "simplicity" and "clarity." The idea of evolution was invented in response to supposed philosophical requirements and at first existed as a completely abstract concept, without any reference to reality.

Charles Darwin's grandfather, Erasmus Darwin, promoted the idea that the power of generation has a much greater creative potential than the faculty of reason, which people brag about. He was the first to propose, in his book *Zoonomia* (1794), the idea that all animals descended from a "single filament" by natural generation, without participation of any intellect. Nevertheless, Erasmus Darwin believed that the power of generation, which gave origin to all species, ultimately comes from the "Great Architect"—God. According to Erasmus Darwin, God is more noble and perfect in our eyes precisely because He does not create effects themselves but rather "causes of the effects." In other words, He employs secondary causes to bring out the effects and this fact reveals more of His greatness.

We can say that Erasmus Darwin was the first to present a theistic version of evolution, which in its theological layer does not differ from contemporary theistic evolution. In 1817, Erasmus Darwin's book was placed on the papal Index of Forbidden Books, in a more severe category, i.e., books that contain so many errors that they are not suitable for correction. Therefore, the first Church condemnation of the naturalistic idea of the origin of species took place long before Charles Darwin. Moreover, it was not about evolution that excludes the existence of God, but the "theistic" form of evolution, somehow initiated or influenced by God.

But today it is argued that Charles Darwin's On the Origin of Species *was never placed on the Papal Index...*

The Index was governed by certain rules. One of them was that it generally included books by Catholic authors. Works by non-Catholic authors that touched on religious issues were considered heretical by default. For example, when a Protestant author wrote a work that considered theological issues, this work could be deemed free from errors only if the Congregation of the Index (or another office) issued a positive opinion stating that the work in question contains no errors.

Charles Darwin's work *On the Origin of Species* was considered scientific and thus avoided any serious ecclesiastical scrutiny and therefore judgment. But in fact, Darwin dealt with religion. For example, Polish philosopher Prof. Kazimierz Jodkowski lists ninety-two theological claims in Darwin's main book. Therefore, we can say it was condemned *a priori* and the Congregation of the Index did not have to confirm this by a separate act. The fact that *Zoonomia* was included in the Index shows how much the Church wanted to emphasize her opposition to the idea of biological macroevolution. I think, therefore, that the argument from "non-condemnation of Darwin" stems mostly from ignorance of this historical reality.

Were there any other pioneers of evolution before Charles Darwin, apart from Erasmus Darwin?

A few other naturalists of lesser rank published evolutionary ideas in England at the beginning of the nineteenth century,[1] but they were met with opposition from the authorities of the Church of England, which at that time *de facto* meant opposition from the state, the universities, and the entire system. So, their ideas did not reach a general audience. It was different, though, with Robert Chambers, who published anonymously his book in 1844.[2] Chambers' book prepared the way for Darwin by familiarizing a broad audience with "transformistic ideas."

A more famous evolutionist was Jean-Baptiste Lamarck, who lived and worked in France.[3] Lamarck's ideas sometimes become a target of ridicule because he believed, for instance, that the giraffe has a long neck because when there was a shortage of food in the low parts of the trees, it had to stretch out and in this way the neck became longer over the centuries. Similarly, reptiles began to fly because when catching prey they had to make longer and longer jumps until finally they became birds.

Lamarck's ideas were discredited by much later discoveries in genetics, which proved that acquired characteristics are not inherited. Today, when we know that genes do not contain all the information necessary for the functioning of an organism, it is not excluded that besides some genetic mutations that are inherited there are some other acquired characteristics that may be inherited, as well. Hence, even serious biologists try to rehash Lamarck's theory. As far as I am concerned, this is just another confirmation that evolutionists are wandering around in circles.

How was Lamarck's theory received in France?

Lamarck, of course, assured the public that the entire evolutionary process was not accidental, that it would not have been possible without the creative action of God. But Lamarck lost the battle on the field of science. His ideas were opposed by Georges Cuvier, a great geologist and certainly one of the greatest naturalists of the Enlightenment. There were several academic debates between evolutionists on one side and supporters of the fixity of species on the other, during which Cuvier crushed Lamarck's ideas not only with his arguments, but also with his academic gravitas. Cuvier claimed that the fossils known to us do not represent lines of development from one species to another; instead, each of them represents a distinct, fully developed organism. There are no transitional forms between them. The fixity of types and the lack of intermediate species became a convincing argument against species transformism. Evolutionary ideas subsided in France for a long time.

So, can we say that evolutionary ideas lost the debate in the scientific community in the first phase?

It is worth noting the context of these early debates on evolution. Early evolutionists were usually not professional naturalists. Lamarck started out as a soldier, although he gradually gained a scientific education and reputation. Erasmus Darwin was a physician more often recognized as a ladies' man and a poet. He was considered a good doctor, although in his time medicine was more quackery than real science. It was said of him that he had served Bacchus and Venus all his life, but due to his deteriorating health he could no longer serve two masters. So, he abandoned Bacchus, but remained faithful to Venus until the end of his life.

Charles Darwin was more of a theologian than a naturalist, from a strictly professional point of view. He did not complete the medical studies to which he was sent by his father. His family, crushed by this failure, wanted to make him an Anglican minister. Eventually, he managed to finish his basic studies in liberal arts with a bachelor's degree, and this was his only formal education. Notice, then, who created the theory considered to be the basis of modern science: soldiers, poets, humanists.

These amateurs were opposed by the greatest researchers of the time—Cuvier in France and Charles Lyell in England. These scientists knew much better the facts that Lamarck and Erasmus Darwin referred to. Nevertheless, their knowledge of these facts did not induce them to accept biological macroevolution. From this we can conclude that evolution was imposed on the data of nature as a philosophical and theological idea, rather than a result of any real discovery.

If the basic idea of evolution was formulated already before Charles Darwin, then why is he considered the inventor of evolution? What new thing did he bring to the debate on evolution?

Charles Darwin was the first to propose a specific mechanism according to which nature might create new organs and new species. Previously, scholars only imagined that some creatures transformed into others. Lamarck's explanation was unconvincing and still required some divine plan or general providence. However, if something is a product of nature alone, i.e., if it came about thanks to some material processes, then there must be a natural law that is responsible for that effect.

Darwin presented such a law—natural selection operating on random variation. This mechanism does not require any intelligent guidance. Blind nature by herself, operating according to its own rules, creates biological novelties. Therefore, it was only Darwin who managed to do what his predecessors dreamed of—explain the diversity of biological forms without referring to divine causality, or any design, or purposeful action.

I suppose that there are many objections that may be leveled against Darwin's theory. What would you consider the greatest problem of his proposed "mechanism" of evolution?

Indeed, each step of Darwin's reasoning is doubtful. But what I would consider the greatest obstacle to his mechanism is the problem of the emergence of biological novelties, i.e., new organs performing new functions. Natural selection preserves only those biological structures that give an advantage in the organism's struggle for existence. If a novelty does not give an advantage, then it is invisible to natural selection. If it reduces fitness, natural selection will eliminate such novelty. Therefore, a new feature that reduces the chances of winning the struggle for existence will lead to the elimination of the individual who has this feature.

Now let us ask, What really gives an advantage in the struggle for life? If there are such things at all, these are complete, ready-made organs or functions. On a microscale, these are completely new proteins perform-ing new functions, while on a macroscale, these are organs such as wings, lungs, gills, bat sonar, etc. But such organs cannot arise in one evolution-ary step, i.e., within one generation, because they would require changes in an organism that are too radical. Darwin also knew this, which is why he kept repeating that evolution can only work in tiny steps, gradually modifying organisms.

Note, however, that a small step does not produce a new organ, and therefore does not give any advantage in the struggle for life. Who needs half of a wing, one feather, an eye that has no nerves and cannot see? If such defective organs appear by chance, natural selection will eliminate these organisms first. Moreover, we do observe natural selection in nature when, for example, disabled or mutated individuals die prematurely or fall victim to predators. Therefore, evolution could either work in large steps or not at all.

If, however, a reptile were to suddenly grow feathers and wings and begin to fly, we would consider this nothing less than a miracle. Evolution does not work through miracles and nature does not take great leaps and therefore, according to the principles adopted by Darwin, it cannot produce biological novelties.

How do evolutionists respond to this?

Of course, evolutionists believe that there is no problem here, because these large leaps are achieved by means of a series of small "islands" on which Darwin's mechanism blindly jumps until it has crossed the whole ocean. But notice that this answer does not solve the difficulty.

The objection is that it is impossible to go from an organ performing function A to an organ performing function B in a single evolutionary step. Evolutionists say that although it is impossible to go from A to B in a single step, there is an organ performing function C between A and B. It is not possible to go from A to B directly, but it is possible to go from A to C and then from C to B.

But this argument, instead of resolving the problem, rather multiplies it. Why would going from A to C, or from C to B, be any easier than going straight from A to B? After all, a new function must emerge in one of these transitions, so multiplying the intermediate stages does nothing but hide the problem under one of these transitions. And adding more intermediates ends up in an infinite regress. But if the series of intermediate organs and functions is infinite, then evolution will never get through it. This is why the transition from A to B is still impossible by the Darwinian mechanism.

Father, you are talking about a certain pattern of thought that poses a problem for evolution. How does this relate to the real world of biology? Are there any experiments confirming this difficulty?

All experiments focusing on the "creative capacities" of the neo-Darwinian mechanism show that one evolutionary step (i.e., a single or double coordinated mutation) cannot produce any relevant biological novelty. Even transforming one protein in such a way that it performs the function of another, a very similar protein requires at least seven coordinated mutations.

Indeed, the neo-Darwinian mechanism can sometimes go a few steps forward, when the destruction or weakening of some function (which can be done in one step) accidentally gives an advantage in the struggle for life. Michael Behe compares this to the situation of preparing a car for a drag race. What can be done to make a car go faster? You can break off side mirrors which will reduce air friction, and pull out the passenger seats and remove the spare wheel in order to make the car lighter. Then, such a modified car would win a short race on a straight street. In this case, damaging the car had a positive effect.

But we know that such a car is not suitable for normal use. The damage to the mirrors is not a creation of any new function, it is a breaking of the car in such a way that in very specific circumstances it has a short-term advantage over the "healthy" car. But in a normal situation this "improvement" would be a cause of disaster. The car would be quickly eliminated from the road, either by natural selection (accident) or by guided selection (Police). And such is the nature of random mutations.

What I am saying has been confirmed by numerous experiments and described in scientific language in professional journals. For details I refer you to the publications by the Biologic Institute (the journal *Bio-Complexity*) and also to the more popular book by Michael Behe, *The Edge of Evolution*.[4]

Does it mean that the Darwinian mechanism does not explain anything?

To some extent, it explains microevolution, that is, the appearance of minor changes in populations of individuals of the same species. In microevolution no new organs are created that perform new functions. Only organs, or any biological functional entities, that already exist undergo minor modifications.

For example, the Galápagos Islands are home to finches—birds that Darwin classified during his journey around the world. Finches vary in the size of their beaks. Today, we know exactly how microevolution occurs in these birds. During dry years birds with slightly larger beaks (on average by 0.5 mm) begin to dominate the population, because they are able to extract food from dried-out soil. When humidity returns to normal, the proportion of finches with small and large beaks also returns to normal. Nothing new emerged in the population, only the proportions of individuals with smaller and larger beaks were changing. It does not follow that beaks would change into other organs nor that climate pressures produced the first beaks. Microevolution is therefore a fact. It is just that microevolution is not controversial.

Everyone acknowledges the fact of microevolution, such as malaria parasites gaining resistance to antibiotics due to genetic mutations. The mutated malaria parasite will be selected in an environment in which antibiotics are present; therefore, in such circumstances it will come to dominate the population. In this case, the neo-Darwinian mechanism (random genetic mutations and natural selection) seems to explain the adaptive capacity of organisms. Again, it does not follow that the same mechanism led to the emergence of the malaria parasite, nor that the parasite will change into a different creature. However, modern biology has doubts whether even this simple mechanism is completely random.[5] Ultimately, it may turn out that pure neo-Darwinism does not explain even these basic examples of microevolution.

Does the so-called "forbidden archaeology" contribute anything to the critique of Darwinism?

In my research and in what I say here I am careful to distinguish science based on actual data from charlatanism, ideology, or pseudoscientific sectarianism. And this applies to both sides of the evolution debate. "Forbidden archaeology" is a term that usually describes fictitious paleontological data that is supposed to prove the most incredible theses. It seems that the term was popularized by Richard L. Thompson and Michael Cremo, who published a book[6] in which they argued that humans lived hundreds of millions of years ago, at the same time as dinosaurs.

Scientists do not take this seriously. I think the authors themselves tried to gain popularity by originality. Such revelations would be irrelevant for our debate if it were not for the fact that honest criticism of evolution can be easily ridiculed by associating it with pseudoscience. I am not saying that there is no such thing as pseudoscience. But at the same time I am saying that pseudoscience is only a negligible margin of scientific criticism of the Darwinian theory. Unfortunately, this distinction is typically not recognized by the propagators of theistic evolutionism. They quite easily lump together disparate ideas such as "forbidden archeology," Young Earth Creationism, and intelligent design, and label all of them as pseudoscience. *Distinguere sapienti est.*

What is the difficulty with the so-called Cambrian explosion of life?

The fact of the Cambrian explosion is a difficulty for the Darwinian explanation of the biological diversity of creatures. As we know, Darwin presented the history of life in the form of a great tree. In the beginning there was only a "trunk," which symbolizes some first organisms, from which gradually emerged the next ones, up to the ones we know today.

If this story were true, then in the fossil record, i.e., the succession of fossils, we should observe these gradual transitions represented by millions of intermediate linking forms. In fact, all organisms (with the possible exception of the first and the last one) would be only intermediate forms. But in the fossil record we do not have these links. We find fully developed organisms adapted to their conditions, which do not change significantly over the millions of years of their existence.

However, what is most striking about the fossil record is the fact that almost all biological types appeared at a geological moment, about 540 million years ago, in the era called the Cambrian. Recent studies have narrowed the geological window of the Cambrian explosion to 10 million years. The problem is that 10 million years is at the limit of the resolution of the "clock" we use to study fossils. In other words, it is hard to tell how long an event lasted if it lasted less than 10 million years. It seems therefore that all this diversity of life could have appeared over 10 million years, or in a single moment. This is irrelevant to the evolution debate, because even 10 million years is far too short a time for Darwinism to produce all that diversity. The general conclusion from the fossil record is that large groups of organisms appear suddenly, as if from nowhere. And this positively rules out the Darwinian tree of life. This tree has neither trunk nor branches.

I suspect that Darwin's supporters have an answer to this problem.

Darwin already knew that the fossil record does not represent a pattern of gradual transitions. He wrote about the large number of missing links. In response, however, he claimed that the fossil record is simply incomplete and that in the future we would find these missing links. He also thought that they were buried, probably under the ocean floors. This was quite clever, because in his time it was impossible to study what was under the ocean floors. However, geology has made enormous progress since then. It turned out that there are no connecting forms under the ocean floors, and the oldest fossil layers are more common on land. So, no links were found.

But paleontologists announce the discovery of a new link almost every day. So how is that?

Yeah, it's pretty funny. Listen to these paleontologists. If you ask them about missing links, they'll tell you that basically nothing is missing. There are some minor holes, but basically we have a complete evolutionary chain of all the important groups of species. And then the next day you will hear that they just found the "last missing link" that paleontologists have been looking for for years. But how can you find missing links if there are no missing links?

The problem is that paleontologists only admit that they didn't have a link when they find something new that can be presented as a link. Soon after, it turns out that it is not a link at all, but a new unique species that was not known to have existed. But you will not hear about that from the popular media.

Can we say that the gaps in the fossil record are gradually being filled in by new discoveries?

I have already pointed out a more fundamental problem with all evolutionary thinking. If you believe that similarities between organisms indicate their common ancestry, then you will always prove common ancestry. After all, every newly discovered organism bears some similarity to other organisms.

If a fossil is incomplete, it is "reconstructed" in such a way that it fits the evolutionary scheme. Thus, every discovery confirms the theory. The lack of discoveries also confirms it, because it only proves that fossils have not been preserved. It is like blind fate, a self-fulfilling prophecy. One of the official slogans of the Communist Party in Poland was "Once we have taken over the government, we will never let it go." Here a paraphrase fits: "Once we have accepted the theory, we will never give it up!"[7]

Using the evolutionary *a priori* paradigm, you can say that there is no discovery that could harm the theory. You don't need to be a genius in methodology to see that something is off here. But the materialist paradigm is so strong in contemporary culture that it easily absolves investigators from even the most obvious methodological errors.

Does this mean that paleontologists do not see a problem even with the Cambrian explosion?

The Cambrian explosion is a different issue from the lack of intermediate forms. The explosion falsifies the existence of the tree of life. In other words, if we agree that such a paleontological fact occurred, the Darwinian tree of life can no longer be maintained.

Evolutionists once tried to dispute the Cambrian explosion by saying, for example, that organisms showing the gradual development of Cambrian forms have not been preserved in the fossil record. But this argument failed. Since we have fossils of early single-celled organisms (prokaryotes) from 2–3 billion years ago, as well as the perfectly preserved fossils of the so-called Ediacaran biota from the Precambrian period, it is not justified to assume that intermediate forms which could prove the evolutionary origin of the Cambrian fossils once existed but were not preserved.

Again, for the details of this debate, I refer you to the works of scientists. I especially recommend Stephen Meyer's book, *Darwin's Doubt*,[8] and an earlier popular film, *Darwin's Dilemma*.[9]

What did the discovery of DNA change in the debate over evolution?

In 1953, James D. Watson and Francis Crick discovered that the process of life is based on information recorded in the cell nucleus in the form of a sequence of specific chemical compounds. They immediately noticed that this sequence resembles the arrangement of letters in a sentence. Their explanation was widely accepted and is still valid today.

The nucleotide bases in the double helix of DNA constitute a kind of encoded sentence. What is important is that their sequence (the order of base pairs in the chain) is not determined by any chemical, physical, or structural law, yet it is this specific sequence that determines the functions performed by the cell. The informational nature of DNA typically is not a subject of controversy. The point of contention is the source of this information.

According to the neo-Darwinian theory, the information necessary for the functioning of various organisms was formed through random mutations over millions of years. This is equivalent to saying that a multivolume encyclopedia can be created by a random generation of letters. Some believe it, others do not.

The alternative theory—the theory of intelligent design—claims that the creation of that huge quantity of coherent information required the action of an intellect or a force capable of selecting for a specific purpose. The selection of characters in the code was therefore not random, which of course does not rule out the possibility that random mutations could later change or even destroy this original design to some extent.

So, we have two competing theories aimed at explaining the origin of the current genomes and the genetic code itself. But the real challenge for materialistic evolutionism is the question of where the first information necessary to build the first organism came from. Before life appeared, there was no inheritance of random mutations, because there was no life cycle nor information that could mutate. Therefore, for neo-Darwinism to generate more information, the first information must already exist.

Moreover, there must be cellular machinery that reads this information. Cellular machinery and DNA constitute irreducible complexity, which means that in a living organism one cannot function without the other. Evolutionary variants of the origin of life would have to explain how both of these realities arose simultaneously. Thus, the discovery of DNA revealed many serious shortcomings in Darwinian explanations.

I understand that materialistic evolutionism encounters an obstacle in the form of the Rubicon of life: The neo-Darwinian mechanism can operate when there are already reproducing organisms. But it does not explain the origin of the first organism. How do evolutionists respond to this problem?

The mechanism that Darwin thought of cannot operate unless there is already a process of life. Darwin and subsequent evolutionists said that the first life arose spontaneously in some warm swamp filled with basic chemical compounds. This explanation may have seemed plausible when a cell was considered a "lump of protoplasm," something like "squishies" (jelly toys) for children made of a homogeneous substance.

All of evolutionary thinking was backed by the assumption that the first life was "very simple," that one could go from non-life to life in a few simple steps. But today we know that there is no such thing as "simple life." In fact, the smaller the structures we study, the more complex the phenomena we encounter. The cell, which is the basis of all life, turned out to be an entire microcosm of dependencies and functions.

So, there is an incredible complexity at the bottom, which then seems simplified when we move to the macroscale. Organs observed with the naked eye (such as wings and hands) or internal organs (such as the heart, lungs, and muscles) are relatively simple in their basic structure and function, compared to the biological devices we encounter at the cellular level. But it is this basic level of cellular life that decides the "to be or not to be" of organs on a macroscale.

This turns the entire evolutionary logic upside down: First comes complexity, not simplicity. In fact, we have no credible theory of the origin of life. This problem has been described in detail by Stephen Meyer in his book *Signature in the Cell*.[10] For more details, I refer you to this publication, as well as to online lectures by a leading chemist, James M. Tour.

So far, we have been focusing on the scientific problems of Darwinism. What does theology say about Darwinism? Could God have used evolution to bring about all the diverse forms of life?

If God used evolution, then it is not an unguided process, so it certainly is not Darwinian evolution. However, Darwinian evolution is the only kind that is widely accepted in science. (A few alternatives have been proposed, but none of them have really taken off.) Therefore, even if we accept that God could have used evolution, that does not solve the problem of the conflict between Darwinism and Catholic theology.

But I think there is another, perhaps even bigger, problem with the formulation "God could have used evolution." This is its hypothetical character. Ultimately, God can do anything, so such a statement does not contribute anything to the debate. The question is not what *God could have done*, but what He *actually did*, and whether we have any method to establish it.

Should a theologian engage in the scientific debates?

I am a theologian and a philosopher, so the scientific debate on evolution is only a secondary subject of my interests. Therefore, when we discuss the issues of natural science, I consistently refer to the professional scientific literature for the details.

At the same time, I believe that it is the duty of every theologian and philosopher to know the basic scientific arguments "for" and "against" evolution, especially if the theologian makes a public statement. When you ask the average theologian why he accepts the evolutionary origin of species, the standard answer will be something like this: "Biology has proven evolution. I am not a scientist, so I will not argue with the achievements of biology. From the point of view of faith, everything is God's creation, anyway. If God wanted, He could have used evolution."

Notice that this theologian claims that he knows next to nothing about biology, and yet he accepts evolution without reservation owing to the alleged biological evidence. But ignorance can only justify a suspension of judgment. Moreover, those theologians who "don't know biology" often turn out to be very zealous in defending the evolutionary paradigm in theology.

To me, this kind of attitude seems irrational. If a given theologian does not know biology, perhaps he should first check whether the alleged evidence for evolution is really so robust that theology needs to be modified in light of it. Moreover, this theologian does not know that there is a whole set of powerful biological arguments against macroevolution. Therefore, ignorance essentially prevents him from forming an accurate judgment on the scientific part of the story. As a result, the theologian is unable to defend the faith against the unjustified claims of materialists and atheists, who are very good at using advanced science to popularize their materialistic worldview.

Are there any arguments against evolution outside biology? For example, philosophical ones?

Of course. I have already mentioned theological arguments. These are primarily the revealed truths about the origin of species and especially human beings. This is the sacred Tradition of the Church, which excludes Darwinian evolution, and at the same time unequivocally supports the formation of the first human body by the power of God directly from the dust of the earth.

There are many philosophical arguments against biological macroevolution, even in its theistic version. However, before we talk about them, I must first point out that philosophical and theological arguments are different in nature from scientific ones. The latter are based on experiments and empirical observation.

In contrast, theological arguments rely on the premises derived from divine revelation, which are simply accepted by faith. Philosophy, in turn, is based on the contemplation of reality (being as being) perceived at the most general (abstract) level. In Christianity, there are two philosophical traditions: one is the Platonic-Augustinian, and the other is the Aristotelian-Thomistic. Both traditions of Christian philosophy exclude theistic evolution, which is the idea that God did not create species directly, but instead used secondary causes such as material evolutionary processes.

Why does Christian philosophy exclude theistic evolution?

I do not want to go too deeply into the difficult language of classical metaphysics, so I will cite only one argument for each tradition.

Aristotelian-Thomistic metaphysics says that every being consists of an essence, or *what a thing is*, and accidents, or *what it has*. This can also be understood based on the distinction between what something is and what something is like. The essence of a cat is its "catness," which is its specific feline nature. But when we see a cat, we do not see its essence, but rather its accidents. We see that it is white, has black eyes, a long tail, and other features. Based on the perception of these features, our intellect forms the concept of a cat, and thereby recognizes what it is.

Thanks to the faculty of forming abstract concepts, when we see another cat—a black one with green eyes and a short tail, we know that it is also a cat. We know this, even though the second cat differs from the first one in many of its features. Therefore, the essence of a cat is not our invention; it must somehow exist in these individual cats. We see that features do not change the essence of a given animal. A cat remains a cat, regardless of the color and length of its fur or its many other features.

But according to Darwin's theory, it is precisely changes in features (such as genetic mutations)—which are always at the level of accidental changes—which lead to the creation of new species, i.e., completely new beings or natures. In this way, a reptile is supposed to change into a bird and a monkey into a human.

According to classical metaphysics, the Darwinian mechanism can only change accidents (features), not the essence of a being. Thus, it does not matter how long evolution accumulates accidental changes, it will never produce a new substance. Therefore, the transformation of species through biological evolution is impossible.

In fact, Darwin's theory implies that substances or natures do not exist, which means there is no such thing as a cat nature, monkey nature, or human nature. There are only individual beings which are just links on the long path of evolution. Of course, you can immediately see the consequences of such thinking, for example, for morality.

We will return to moral issues in our conversation. What is the second philosophical argument against evolution—the one stemming from the Platonic-Augustinian tradition?

According to the Platonizing Christian tradition, creation occurred according to divine ideas. That is, in God, in the divine mind, there are ideas of all the beings that can be thought of, and more. Certain ideas are also in our minds. For example, ideas of material objects, such as a car or a telephone, ideas of animals—a horse, an elephant, or a giraffe—as well as ideas of general concepts that cannot be visualized, such as the idea of justice or mercy. Of course, ideas are in our minds differently than in God. In ours, they are like apples in a basket, while ideas in God are not something separate from God himself. They belong to His essence, that is, they are one with God himself. Divine ideas are infinitely more distinct and clear, and are in God from eternity, all at once and unchanging. We, on the other hand, must acquire ideas through experience or enlightenment. We learn them gradually.

Christian tradition says that the world was created in accordance with ideas inherent in the divine intellect. These ideas are said to be "exemplars" for creation, meaning that individual elements of the world materially realize the concept contained in these ideas. For example, horses, cats, dogs, and lizards were created in accordance with the original divine idea of each kind which God impressed, as it were, upon matter in the form of a new substantial form which, in combination with matter, created a new being, that is, a new species. But in Darwin's theory, properly understood, there actually are no "species"; there are only individuals that constitute transitional links. Therefore, according to his theory, divine ideas were not realized in creation because everything is in a constant flux.

Do such arguments even matter to anyone today? After all, they concern only some particular philosophical concepts. What relevance do they have to the debate on evolution?

First of all, these arguments are not intended to replace theological or scientific arguments, which we have discussed at length. They are also not the only philosophical arguments that exclude theistic or Darwinian evolution. Every theologian and philosopher must be aware that the debate over evolution is not just a matter for biologists, but has enormous significance within philosophical and theological discussions. The problem is that if we accept "Darwinian metaphysics," we must abandon classical metaphysics. For Christian theologians and philosophers this is not an option, because the main dogmas of the Church were formulated in terms of classical metaphysics.

Could you give some examples?

Trinitarian dogmas are based on the metaphysical concepts of nature, essence, and person. Similarly, Christological dogmas are based on the concepts of nature and person. In turn, the Eucharistic dogma is based on the concepts of substance and accidents, which I mentioned. By accepting Darwin's theory, we must abandon these concepts. But if we abandon classical metaphysics, we do not understand our faith; we cannot define it or convey it in an objective language. Surely, it is true that we do not believe in the dogmatic formulas themselves, but in the reality that these formulas describe. Still, it is also true that despite many attempts, no one has succeeded in describing these realities in any language other than the language of classical metaphysics.

Therefore, by abandoning this metaphysics, we deprive ourselves of the rational explanation for our faith, thus turning faith into fideism. I think that unfortunately very few theologians realize that the effects of Darwinism go that deep. And this is not about some "ideological" interpretations of Darwin's theory, but about its very core—the erroneous metaphysics on which this theory is based.

How does theistic evolution relate to the Christian understanding of creation?

I will draw attention to three fundamental points in which theistic evolutionism reinterprets the classical Christian doctrine of creation.

First, according to this view, after matter emerged from nothingness (a moment usually identified with the Big Bang), God never again acted directly in the history of the formation of the universe. However, according to the Bible, the Church Fathers and holy Doctors, and essentially all theology until the late nineteenth century, God shaped the world by his supernatural action over a period of time that the Holy Scriptures call six days.

Second, according to theistic evolution, God uses creation when creating. In other words, God somehow transfers the power of creation upon His creation. In the classical Christian approach, however, no creature can serve God as an active helper in creation. The idea that God conferred the power of creation upon the world was always considered a grave heresy.

Finally, according to theistic evolution, creation never ended. Proponents of this view accept the concept of continuous creation (*creatio continua*). In contrast, in the classical Christian view, creation ended once and for all with the creation of man. After the end of creation, God sustains things in existence (*conservatio rerum*), but He does not create anything new. Hence, theistic evolution does not recognize the Sabbath of the Lord which means it is at odds with the Christian tradition and the Bible.

You can clearly see that theistic evolution substantially modifies the classical Christian understanding of creation. It is therefore not an orthodox view.

7. Intelligent Design—Fact vs. Fiction

I read in several books and papers that the theory of intelligent design (ID) is entirely discredited and there is nothing to talk about. Has the theory really come to a dead end?

Certainly, I have also heard this thesis many times. Nevertheless, I disagree. Authors who criticize intelligent design (ID) do not support their critique with proportionally strong arguments. Based on what they say, I must conclude that many of them have not read the books presenting arguments for intelligent design. Apparently, they only know a caricature of the theory as it is reproduced in unfavorable literature.

Quite often, the critics of intelligent design misrepresent the case for ID and then rebuke the made-up argument. A good example is their repeatedly saying that intelligent design claims that "life is too complicated to come about spontaneously." To disprove this allegedly core ID thesis, they provide a number of examples of how complex structures emerge in nature spontaneously—by chance, or by necessity (laws of nature), or a combination of both chance and necessity.

This is a typical instance of a "straw man" fallacy. ID does not say that some structures are too complex to arise randomly, but that they are irreducibly complex and—as such—they cannot arise according to the neo-Darwinian mechanism of genetic mutation and natural selection. Irreducible complexity does not have to be extremely complex (even if it usually is), but it must be both *irreducible* and *complex*.

We will return to irreducible complexity later. Why do you think so many theologians and philosophers disregard the theory of intelligent design?

Some of the critics just repeat easy theses without testing their credibility. Again, it is more a problem of "bad knowledge" than "bad will." Many of them are good people who simply do not have the time to look up source literature. Their fault is that they line up with the wind without having checked if the wind pushes them in a correct direction. I am speaking mainly about theologians and philosophers working at Catholic institutions. Besides them, there is a significant number of people whom I would call "vulgar Darwinists." These are mostly atheists working at secular universities who devote their lives to promoting Darwinian evolution. Debating them is rather futile. The dedication to the Darwinian paradigm that you see among the "vulgar Darwinists" is amazing. Their belief in evolution is something more than just a conviction that a certain scientific concept is true. Without exaggeration, we can say that Darwin's theory plays in their lives a role akin to a religion.

I think that, at least for now, it is more important to discuss the issue in the Church herself, because only after we clarify matters among believers can our mission be successfully extended to pagans and atheists. Only then will the Church shine again with the full splendor of truth. As John Paul II put it: "The truth cannot impose itself except by virtue of its own truth."

Are there any particular reasons for the negative reception of intelligent design among Catholic scholars, especially in America?

Not surprisingly, the understanding of the controversy over evolution is generally better in America than in other countries, including Poland. Not only is the English-speaking culture the home for these ideas, but also academic abstract thinking is highly developed in America. One thing that seems to be particular to America is a phenomenon popular in some philosophical circles that I would call the "reduction of science to philosophy."

What does that phrase mean?

Let me explain it in greater detail, because, I think, this is the key to understanding why some Catholic scholars reject the theory of intelligent design.

Many conservative scholars study classical Christian philosophy and try to promote it in the contemporary intellectual culture. This project, generally speaking, is important and praiseworthy. It stems from the conviction that ancient and medieval sages, such as Aristotle and Thomas Aquinas, discovered some eternal truths that are valid for any culture at any time.

This approach, however, oftentimes is not balanced by a proper discrimination of what is really eternal and permanent in the teachings of such sages from what is just accidental to the times and contexts in which they lived. I agree that the theological and strictly philosophical ideas are permanent in the sense that they do not evolve, they do not change their meaning along with evolving culture. Once true, they are true forever. But the same principle does not apply to the ancient or medieval scientific concepts. Many of them are indefensible in our day.

So, you are saying that the scientific ideas of the ancient philosophers may be outdated, while their strictly philosophical concepts are permanent. The discrimination between the two makes it possible to properly grasp the relationship between classical philosophy and modern science. How does it affect our understanding of classical philosophy?

Modern science generally eliminated final and formal causality from the explanations of natural phenomena and focused entirely on efficient causality, which in turn was reduced to merely material causality. And this is a great offence for traditionally-minded philosophers. Due to their eagerness to rebuild objective and realistic discourse in all human intellectual engagements, they tend to criticize science for not employing metaphysical terms, such as final and formal causation.

This is, in a way, a violation of the proper and just autonomy of modern science. Scientists who work with modern equipment, study proteins, or compare different chemical reactions in cells are encouraged by these philosophers to seek final and formal causes, and if they fail to do so, they are accused of "materialistic" or "mechanistic reductionism."

Such "judgmental" philosophers tend to forget that modern natural science is autonomous due to a different purpose, method, and object from, let's say, Aristotle's natural philosophy. In fact, traditional philosophers who try to introduce strictly philosophical notions into natural science do not counter reductionism. Instead, they propose their own type of reductionism, which is the reduction of science to philosophy.

From what I am saying it does not follow that no hierarchy between different disciplines exists. I agree with the classical thinkers that theology is the queen of the sciences and that philosophy is nobler than natural science because it asks about the *archē*—the ultimate source or cause of everything—while science focuses on particulars and pragmatic goals. However, precisely because of the objective hierarchy of the disciplines, the autonomy of natural science should be respected.

Science does not need to resort to strictly philosophical notions, such as formal and final causality, in order to achieve its proper end, which is to explain material reality pursuant to its own method. It does not follow that final or formal causes do not exist. It only means that they are outside of natural science.

How does this specific kind of reductionism—science to philosophy —affect the discussion about intelligent design?

Intelligent design claims that design can be detected in nature by means of strictly scientific methods. This means that the partially philosophical idea of design may be discovered and demonstrated without employing specifically philosophical tools. Some philosophers—especially, if they do not understand intelligent design arguments correctly—feel endangered. Not only do some scientists claim to have discovered the idea of design in nature, they also boldly maintain that they can do it without resorting to philosophy.

The philosophers worry that in this scenario natural philosophy is not needed anymore. Moreover, they may suspect that scientists would not need formal and final causes at all, because they discovered something that simulates those ideas through intelligent design. So, in the minds of those philosophers, ID reinforces the type of reductionism that they see in natural science and try to combat. Intelligent design becomes a hostile idea, not because it denies finality, but because it introduces an incomplete (in their view) idea of finality without reference to the great minds and their great arguments from the past.

Paradoxically, those traditionally minded philosophers criticize ID along with overt materialists. The difference is that materialists say ID is not science, because it makes philosophical statements, while philosophers of nature say ID is wrong, because it does not make philosophical statements. Materialists assume that the only true and objective knowledge comes from natural science. By removing ID from science, materialists make it "just a philosophy"—in their minds something subjective, vague, and without any real cognitive value. Philosophers, in turn, see ID as mechanistic and reductionist; therefore, for them it is not good philosophy.

Is there any room for intelligent design in this dichotomy of scientific materialism and traditional philosophy of nature?

I think that both parties—materialists, such as the "vulgar Darwinists" and classically minded philosophers of nature—do not see that some biological structures reveal design that is discernable on a natural scientific level without employing typically philosophical arguments. In particular, materialists reject the possibility of discovering design by using the empirical scientific method.

Philosophers, on the other hand, do not realize that the discovery of design in biology with the use of scientific method neither replaces nor diminishes the great philosophical arguments for design and final causality from the past. ID and natural philosophy simply speak about slightly different things from different perspectives.

ID is a new theory that became possible only today, after we have learnt so much more about biology than people of medieval or ancient times could even imagine. No classical philosopher should be surprised that greater understanding of biology results in a new kind of argument for design in nature.

Is intelligent design a new form of creationism?

Creationism in its basic meaning is a belief in separate creation of species. The theory of intelligent design does not require such a belief; therefore, it is not creationism. But it is called creationism by those who are hostile to it. The reason they do so is that it is much harder to rebut the arguments for intelligent design coming from within science than to dismiss "creationism" by opposing it with a "scientific worldview."

Once the debate has been formulated as the simplistic alternative of science versus creationism, it is easy to combat any critique of Darwinism by calling it creationism. This is how many contemporary scholars take a shortcut—instead of responding to the arguments presented by the ID theorists, they call them creationists and thus exempt themselves from the intellectual debate.

Look up, for example, the definition of intelligent design on Wikipedia. Right away, you learn that intelligent design is pseudoscience and a "form of creationism." If this is creationism, you do not need to engage with it, because it is a religious belief which is meaningless in the scientific debate and abandoned long ago in theology.

So, what is the exact difference between intelligent design and creationism?

Intelligent design is a theory based on scientific evidence. It does not resort to any religious texts or tenets of a faith and uses only scientific methodology and natural reasoning. It employs inductive and abductive methods, relies on experiments, and makes an "inference to the best explanation" of the observed evidence. All of these are methods commonly adopted in science.

In contrast, creationism is a religious concept that derives its premises from supernatural revelation. According to creationism, there must have been immediate and supernatural action of God in the origin of species. Intelligent design claims only as much as can be derived from scientific data, namely, that at least some biological structures must have been designed because they could not have emerged by blind biological processes such as random genetic mutations and natural selection.

But if some elements in the biological realm are at least partially designed, there must be a God who designed them. So, intelligent design inevitably leads to a belief in creation.

Even if that were so, it would not take away anything from the scientific character of the theory itself. We need to distinguish between a scientific theory and philosophical or theological implications that follow from the theory. I do not think that the Christian form of belief in creation is necessarily implied by the conviction that there are designed structures in biology.

Theistic evolutionists understand this very well when they say that God did not create species directly but used evolution as a secondary cause. Theistic evolutionists promote a view of evolution as somehow "guided" by God, but at the same time they firmly reject the idea of direct or special creation which is called creationism. Therefore, even according to the opponents of intelligent design, such as theistic evolutionists, it is possible that a designer could have realized his goals without resorting to special creation. This means that theistic evolutionists understand that believing in intelligence as an active cause of creation does not necessarily entail the theological concept of creationism.

By the way, the position of theistic evolutionists ends up in inconsistency. They usually accept Darwinism and claim it is compatible with Christianity, but at the same time they reject intelligent design. In this way they saw off the branch on which they are sitting, because the only form of evolution currently accepted by the Church is evolution guided by God. Darwinism is not that kind of evolution, whereas ID does not exclude "evolution guided by God."

Nevertheless, it seems that the implications of intelligent design are too obvious. The theory carries theological baggage that seems too heavy. Can science bear up under it?

Intelligent design is not the only scientific theory that carries theological baggage. In the early twentieth century, when the Big Bang theory was proposed, some physicists strongly opposed it, because they thought it inevitably led to the acceptance of God and the creation of the universe out of nothing. The seemingly necessary conclusions stemming from the Big Bang theory greatly support the Christian theology of creation—if the universe is not eternal, if it started from a singular point of immense density and temperature, then its creation out of nothing looks like the most reasonable theological explanation.

Today, however, these conclusions are being replaced with the multiverse hypothesis. According to this idea, our world is just one of many worlds which pop up in great explosions and disappear in great collapses in an infinite and eternal multiverse. So, you see that even the most Christian-friendly theory may be countered by the naturalistic and materialistic worldview.

For anyone who believes that the same God who is revealed in the Bible is the one who created the visible universe from nothing, the Big Bang theory is no surprise. After all, this perfect compatibility of the book of nature and the book of the Bible is exactly what we would expect if Christianity were true.

Yet, the obvious compatibility between the cosmological theory and Christian theology does not detract anything from the scientific character of the theory itself. Similarly, the fact that ID harmonizes with the Christian worldview does not diminish its scientific character.

Some scientific theories in the past faced opposition from theologians. They were rejected on the charge that they contradict Christian theology. So, today, how can a believer demand that scientists accept a theory that seems to ruin scientific naturalism?

You are referring to the heliocentric theory. It indeed encountered resistance among some churchmen because they believed that it contradicted both the Bible and the science of their times. However, the objection to heliocentrism was not as strong as it is usually depicted. The Galileo affair was, in a sense, created much later in the Enlightenment. During the nineteenth century, Galileo's case became a leading argument of atheists to promote the idea that faith and science are mutually exclusive.

But it was not Galileo who first proposed heliocentrism. Copernicus did it almost a century earlier. In order to avoid personal engagement in the controversy, Copernicus waited until just before he died to publish his theory. His book, *De Revolutionibus Orbium Coelestium* [On the Revolutions of the Celestial Spheres], provided, among other things, guidelines for the reform of the calendar ordered by Pope Gregory XIII.

Copernicus's work was placed on the Papal Index of Prohibited Books only seventy-three years after its publication. The Index had three categories: books suspended until correction, books banned from teaching, and books condemned. Copernicus's work was placed in the first category, which was the least severe of the three. In the decree announcing the decision, we can read that only very fine (mostly linguistic) corrections are expected, specifically, in order to present the theory as a scientific hypothesis rather than a proven fact.

We need to remember that at the time the competing theory, geocentrism, was a common opinion among both astronomers and theologians. Some of the academic institutions, as well as religious leaders, such as Martin Luther, opposed heliocentrism vehemently. The "condemnation" of Copernicus by the Catholic authorities was no more than just a request to present the argument more accurately, in harmony with the state of knowledge of the time.

Galileo, however, heated up the debate and offended the Pope. Finally, his views were condemned by the Inquisition for contradicting the Bible and natural philosophy (THE science of the time). Thus, it was never a kind of "Bible vs. science" argument. The leading churchmen saw it more as a conflict between two worldviews. Regardless of the fact that over the following two centuries the Church fully accepted heliocentrism, the Galileo affair is still used to heap scorn on religion and specifically on the traditional doctrine of creation.

Does this mean that atheists must come to terms with intelligent design in the same way that the Church accepted heliocentrism?

Scientific evidence is the same for everybody, no matter whether a believer or not. Hence, just as the Church made peace with Copernicus, so too naturalists and atheists should go where the scientific evidence takes them, even if it requires the acceptance of intelligent design.

What we actually see, however, is the opposite of what we would expect. Instead of acknowledging intelligent design, atheists and naturalists challenge the Big Bang theory for philosophical rather than scientific reasons. They try to explain away Big Bang cosmology and the fine tuning of the universe by proposing the multiverse hypothesis. According to them, our universe did not begin in any special way but instead is a part of a cyclic mega-cosmos.

The multiverse hypothesis closely resembles the old pagan idea of the eternal existence of the universe with its infinite circular repetitions of epochs. But there can never be any scientific evidence to support the multiverse because science, by definition, speaks about the universe that can be observed and the only such universe is our own. The multiverse is a purely philosophical (or religious) speculation, disguised in scientific terminology which gives it an appearance of science. Atheists maintain that intelligent design is a religious belief and deny it the status of science, but in fact they propose their own quasi-religious beliefs, such as the multiverse, and claim that this is science.

Then maybe intelligent design, just like the multiverse hypothesis, is a religious theory only pretending to be science?

No, because there is scientific evidence supporting intelligent design. There is no such evidence for the multiverse. Of course, the proponents of the multiverse hypothesis claim that there are multiple data to support it. But look at it from a commonsense perspective. If something is outside of our universe, then, by the very definition of science, we do not have access to it other than through some kind of philosophical speculation.

The error of physicists who believe in a multitude of universes stems from the fact that they are disconnected from what is empirical. They follow advanced theoretical models in physics and after having adopted assumptions within highly abstract cosmological concepts, they conclude that the coherence of these models requires a multitude of universes. Their argument ends up either as circular reasoning or as a completely untested and untestable claim that cannot be considered scientific.

In the twentieth century, mathematical models proved very successful in physics. Many times, equations showed something before the empirical confirmation was obtained. The surprising compatibility of mathematics and physics incited admiration and astonishment among the greatest minds of the last century. But even this method has its limits, and this fact is ignored by the proponents of the multiverse hypothesis. Proposing a multitude of universes simply goes beyond science.

Do you think that intelligent design could ever be embraced by the scientific community?

I think that this theory is at the edge of acceptability for contemporary post-Christian or neo-pagan culture. It is worth noting that thinking according to the intelligent design paradigm is not contradictory to paganism; we know that it was accepted in Antiquity. Platonic ideas are nothing else than "designs" that, although in a defective way, are nevertheless realized in the empirical world. Aristotle's concept of a "form" indicates the intelligibility of being, as well as the intentionality present in it.

Later in the Christian era, this type of thinking about nature constituted the *status quo* to a certain extent. Until the eighteenth century, virtually no one questioned the truth that the world in its various forms reveals the traces of the divine intelligence at work. All of the great founders of modern science, including Copernicus, Galileo, Kepler, and Newton, acknowledged it. This fact by itself proves that pure naturalism is not at all a condition for practicing fruitful science. Science simply must respect the limits of naturalism.

Darwin was the first to deprive science of reference to any intelligence. The naturalists and atheists of the nineteenth century welcomed this new attitude as a sign of the progress of human knowledge. In their view, we do not need to refer the world to any higher causes, because natural processes explain everything we observe. But the examples of Plato and Aristotle show that the classical pagans did not think along these lines at all.

There were materialists similar to Darwin in Antiquity, such as the atomists Democritus and Epicurus, and Empedocles, who developed a primitive theory of natural selection, but their thought did not dominate ancient philosophy to the degree Darwinism has dominated modern biology. In a way, Darwinism resurrects these old pagan ideas, which has led people to think in purely materialistic terms, ruling out the classical reasoning that the world reveals a design that must come from some intelligent being.

From what you say it follows that we live in quite extreme times
when biology is deprived of any reference to intelligence. Can such
an extreme approach prevail, ultimately? If the ancient pagans
discovered design in nature, can the modern pagans completely
exclude it from natural science?

Probably not. It may turn out that Darwinism will share the fate of many
other nineteenth-century "-isms." This is why I said ID was at the edge of
acceptability for modern culture. I do not think that ID would dominate
the biological sciences to the same degree that Darwinism has. But I also
do not think that Darwinism can maintain its monopoly. It is already
falling apart.

Not long from now biology may entertain a plurality of paradigms.
Christians will generally accept intelligent design as a sound scientific
theory which is also compatible with the biblical faith. Militant atheists,
however, will continue to fight and reject any arguments in favor of
design.

Is ID just a remake of the old arguments for the existence of a
Designer?

This is an important question. To understand the theory of intelligent
design we need to see how it relates to the older arguments for the
existence of a Designer. William Dembski—one of intelligent design's
pioneers—distinguishes between the design argument and the design
inference.

The first may be formulated as a syllogism of the following kind: (1)
Everything that is designed must have a designer; (2) the universe has been
designed; therefore, (3) a designer of the universe exists. This is the classical reasoning which finds its roots in pagan philosophers such as Plato
and Aristotle. Christian scholars throughout the centuries also referred
to that reasoning because it is independently confirmed in the Book of
Wisdom (13:1ff) and in the Letter to the Romans (1:19–20). The Book
of Revelation tells us that God—the cause of all things—can be known

by reason alone, just from observation of the universe, without the help of supernatural revelation.

The second type of reasoning—the design inference—implies something different. It may be presented in the form of another syllogism: (1) Everything that has certain characteristics indicating design is actually designed; (2) some elements in the natural world have these characteristics; therefore, (3) some elements of the natural world are designed. Note that this argument does not prove the existence of God, but only the assertion that some structures or events are designed.

In this respect, the theory of intelligent design is a novelty compared to the classical Christian argument. But note that ID does not exclude the old argument. It simply addresses a different question. It remains within science, the natural world, and empirical data.

Can science recognize the presence of design in nature while remaining natural science?

Not only can it, but it constantly does. We can draw on Bill Dembski's examples here. When archeologists find pieces of rocks at an ancient site, they must discern whether these are intelligently designed tools, like axes, arrowheads, or knives, or maybe just random fragments chipped off by wind and water. When a forensic expert investigates the place of death of a man, he must determine whether it was a random event (an accident), or whether it was perhaps caused by an evil designer who worked deliberately to kill his brother. When a computer scientist examines a stream of data, he must be able to distinguish a string of random characters from a program. When a patent office investigates a suspected theft of intellectual property, it similarly must distinguish between accidental convergence of elements and the intentional (designed) ones.

The list of examples could be extended. The claim that there is no place in science for detecting design would destroy quite a large and important part of modern science. Similarly, the theory of ID says that a biologist can distinguish a designed structure or event in biology from the one that is best attributed to chance (such as random genetic mutation) or necessity (such as natural selection).

How is design detected in biology?

The classic argument for intelligent design was formulated by Michael Behe in *Darwin's Black Box*.[1] Behe argues that living organisms contain irreducibly complex systems that could not have arisen by chance and necessity, but required intelligent design.

What is irreducible complexity?

An irreducibly complex system consists of at least a few parts that must all be present in the system at once to allow it to perform its basic function.

Although Behe presented his argument very clearly, there is still a lot of misunderstanding around irreducible complexity. One is the aforementioned claim that, according to ID, some organs are too complicated to arise by chance. But this is not the essence of the ID argument.

We know that even highly complex structures can arise randomly. When you play jackstraws, you throw them on the table randomly, without any plan, but the level of complexity of the arrangement of the straws determines the very quality of the game—the more complexity, the more difficult the game, and the more fun. Similarly rocks undergoing erosion, or crystals in their molecular arrangement, may form highly complex structures thanks to the blind forces of nature. In all these examples, the interplay of chance and necessity produces complexity, but this kind of complexity is not irreducible.

Can you give an example of irreducible complexity?

Prof. Behe refers to a mousetrap to explain irreducible complexity. A standard mousetrap consists of five elements: a platform, a spring, a hammer, a catch, and a holding bar. In order for a mousetrap to work, all five elements must be present at once. If only one of them is absent, the trap does not work. Moreover, each of the elements must fit this particular type of trap. For example, the platform must be rigid enough to stretch the spring. The spring must not be too rigid or too soft, etc.

Behe discusses examples of irreducible complexity in microbiology. He lists, among others, the bacterial flagellum, the cilium, the vesicular transport, and the blood-clotting system. The most apparent example of irreducible complexity for non-biologists is the bacterial flagellum. A flagellum is a filament protruding from the cell membrane outside the bacterium that allows the cell to move through fluid. The flagellum, which rotates at 100,000 rpm, can change its direction in just a quarter of a turn. It is moved by a proton engine that uses electrical bipolarity to create the motion. So, the principle of the operation is similar to the electric engines we are familiar with from daily use.

In addition, this bacterial engine has parts analogous to those of engines made by people: a stator, a rotor, a drive shaft, and bearings. The flagellum is attached to the engine's shaft with a hooked joint. If the joint were absent, the flagellum would rotate along its own axis and would not provide any propulsive power. All these parts must be present simultaneously in order for the flagellum to perform its function. That is irreducible complexity.

Why can't irreducible complexity arise by the Darwinian mechanism of mutation and selection?

Because the Darwinian mechanism works through minute progressive steps that are supposed to eventually take a structure to its function. The instructions for how to build the flagellum are contained in the genetic material of the cell (at least some of them). It is a whole set of coherent information. But, as we said before, one evolutionary step can change only one or two letters of the genetic information at a time. This amount of change is too small by far to produce the whole set of instructions necessary to build the flagellum. Yet, if these mutations do not provide any competitive advantage for the cell, they will be invisible for natural selection and evolution will not move on, leaving them behind.

Let me use an analogy to explain this better. Assume that you bought a TV set and a washing machine. You take the washing machine out of the box and realize that no instruction manual was delivered. Then you take the TV set out of the box and see that someone mistakenly put into the box the instruction manual for the washing machine. Now you have the manual for the washing machine and a TV set without a manual. You need a manual for the TV set. Confident in the creative abilities of your child, you give it the manual for the washing machine and ask the child to change it in a frivolous way. You receive a bunch of meaningless versions of the manual which do not help you to get the TV set to run at all. But the distorted manual does not help you to operate the washing machine either. So, you take away the manual from your child before the text completely loses sense in order to be able to run at least the washing machine.

The same thing happens with random mutations. They do not create new information, so they do not help to develop new organs. Even if they created some part of the information needed to build the flagellum, it would be incomplete. The flagellum would not work and therefore it would not give any survival advantage to these mutations, which would not be selected for. If there is no advantage in a step, the step is not seen by natural selection and it will not be preserved in the evolutionary process.

Moreover, there is a Catch-22 here, because, the higher the mutation rate, the higher the chances that mutations destroy information that is needed for the cell to function normally. If mutations weaken or destroy some essential protein, the organism dies. Thus, a few mutations are not enough to evolve the flagellum, and many mutations most likely would destroy the cell altogether. Hence, the flagellum cannot be built by the Darwinian mechanism.

Do Darwinists have an answer to this problem?

All of us involved in this debate have heard many times that "this problem has already been resolved." It is harder to find out, though, how it has been resolved. In the second edition of his book (published 10 years after the first edition), Michael Behe shows that nobody refuted successfully his anti-Darwinian argument.

Darwinists often claim that the parts of the flagellum were already present in different systems in the cell before they were co-opted to become the flagellum. They fail to mention, however, that about a half of the proteins that are needed to build the flagellum are unique, meaning we do not find them anywhere else in the cell.

So, contrary to what the critics of Behe's argument say, many parts of the flagellum could not have been borrowed from other cellular systems. But a bigger problem is that no one has ever presented a realistic and detailed scenario of how the existing parts would get together and adjust to the new structure and ultimately form the flagellum. The conclusion is that Behe's argument has not been debunked.

Is this not a new version of "god of the gaps" argument? Whenever we do not know an answer on a scientific level we insert God, design, or some unknown power to fill in the holes of our scientific knowledge.

I have heard this objection many times, but, honestly, it is hard for me to respond to it, because I do not even know how to apply it to intelligent design. It is hard to even find any serious examples of such reasoning in the history of our civilization. Sometimes, the example of primitive people is brought up, because they probably thought that thunderstorms were caused by the gods' anger, or that the lunar eclipse meant that the gods were hungry, or something like this.

It must be pointed out, however, that those people did not imagine gods wherever they did not find natural explanations. They simply did not look for natural explanations. The religious interpretation of natural phenomena was satisfactory for them. At some point in Antiquity people started looking for natural explanations of physical phenomena and they found them as far as their tools and methods of observation allowed.

In the medieval epoch—probably the most religious epoch ever—it was believed that, for instance, opium puts one to sleep because it has the "sleep-inducing virtue," and that planets are indestructible, because they are made of a fifth element (or "essence") called "ether." Even though these explanations are mistaken in the light of our modern science, they show that nobody attempted to replace natural explanations with God. In medieval times most people believed that angels move the planets through the sky, but this again was not an insertion of God into the gaps of human knowledge but rather recourse to the best available explanation at the time.

> *But if in medieval times the principle of accepting only material causes had been followed, the progress of science would have been faster. Is this not an argument strongly supporting methodological naturalism?*

If the angels really do not play any role in the revolutions of the planets, then, in this case, the appeal to the law of gravity is the best explanation. So, we follow the principle of the best explanation. Materialists, however, postulate to follow the principle of naturalistic explanations. This would be correct if naturalistic explanations were always the best. But nobody can prove that this is the case. Therefore, I propose we follow the principle of the best explanation rather than decide *a priori* which explanations are the best.

> *We can formulate the charge of "god of the gaps" against the theory of intelligent design in this way: We have a given biological structure. We do not know how it naturally originated, so we conclude it must have been created by a designer.*

This reasoning does not apply to intelligent design. In the case of irreducibly complex systems, we know they cannot arise through the interplay of chance and necessity. We also know from experience that these kinds of systems are built exclusively by intelligent beings like ourselves who design electric motors and mousetraps. Therefore, by applying the analogy to human activity we can properly conclude that irreducibly complex systems in biology are also designed. It is not an argument "from ignorance," because if we didn't know how a molecular system functioned, we could not even say whether or not it was irreducibly complex. It is precisely because we do know the structure and function of a given system that we can infer irreducible complexity and intelligent design. Therefore, it is not an argument from ignorance or "god of the gaps."

> *If there is design, there must be a designer. If there is a designer, he explains everything. There is no scientific curiosity anymore. Instead, we end up in a theory which is no more demonstrable than the ontological argument for the existence of God.*

If there is no designer, the universe is a product of chance and necessity. It is not intelligible. We cannot expect to find anything new or interesting in it. Therefore, there is no sense in pursuing science. You see, the reverse of this kind of reasoning takes us to the same conclusion. In fact, scientific curiosity emerges only if we are convinced that we can discover something intelligible and new. But these kinds of things are produced by minds, not by the blind forces of nature.

> *I encountered an opinion that the unquestionable success of natural sciences took place in Christian Europe because the Europeans started looking at nature as something separate from God and thus "not holy." The desacralization of the physical reality, so to speak, helped to unleash human minds and to investigate the universe. The natural world ceased to be a realm reserved for deities and became accessible to man. Apparently, it was the removal of the gods from the causal explanations that fostered science.*

I think this interpretation is not accurate. It was rather the discovery of God as the Designer and Creator that gave the impetus for scientific progress. Science began to develop in Europe because for centuries it was taught that the world did not come into being out of chaos, but was intelligently designed by a perfect Mind. This premise also constituted a deeper background of scientific research among ancient pagans, such as Aristotle. Undoubtedly, the belief in creation understood as the direct action of God in the universe, which was introduced only by Christianity, was also of no small importance for science, because it confirmed that the design present in nature is not our invention or illusion, but is real.

Today, however, many scientists are atheists who are not hindered in their scientific investigations by the belief that the universe emerged spontaneously from chaos in a grand evolutionary process. It seems therefore that the belief in God and creation is not an indispensable condition for scientific progress.

My answer was aimed at explaining what sparked the emergence and progress of natural science in Europe. And if you consider who created modern science, you will see people believing not only in intelligent design in some generic sense but also in creation understood as direct divine causality in the formation of the universe.

This is even more than ID postulates. The fact that today's atheists thrive on the achievements of the scientists who were believers is a different story. It is understandable that once we have advanced science with a multitude of institutions to propagate it, as well as the demand for new discoveries in industry and business, atheists and agnostics pursue science and contribute to scientific progress. Science has become, so to speak, a separate domain governed by its own rules.

But I dare to say that atheists, who did not create science, have used it many times against humanity and civilization. And, unfortunately, they continue to do so by developing technologies that contradict human dignity such as the ones used in the abortion and contraception industry or some of the embryonic stem cells research. Atheists inherited science as a legacy of Christian civilization. I am afraid, though, that if atheists were the only ones left doing science, within a short time science itself would collapse, assuming optimistically that humanity at large somehow survived.

Let's return to intelligent design. Does ID accept the rest of Darwinism, excluding just a few examples of irreducible complexity?

The proponents of intelligent design do not say that the neo-Darwinian mechanism has no place in biology or that it does not work at all. They claim, however, that it does not explain the origin of complex biochemical systems and for this reason it does not explain the origin of all biodiversity.

The relation between ID and neo-Darwinism in biology may be compared to the relation between Newtonian mechanics and Einstein's general relativity in physics. When Einstein presented his theory, he did not invalidate Newton's discoveries. He only showed that the old theory does not have the same explanatory power. Newtonian mechanics still works, albeit at relatively low speeds and short distances. In order to explain large galactic and intergalactic phenomena, we need a broader theory.

Similarly, neo-Darwinism explains a number of organic changes but in order to explain the origin of biodiversity as a whole, we need a broader theory—intelligent design. You suggested that there are only a few examples of irreducible complexity. But they are found in every cell. We discover more and more irreducibly complex structures and systems.

Moreover, we observe that some irreducibly complex structures constitute larger biological entities, which are also irreducibly complex. We can therefore speak of many overlapping layers of irreducible complexity. And this is the actual way that living beings are built.

Saying that neo-Darwinism does not explain the origin of any complex biological systems sounds like a pretty restrictive statement. Is it not a deadly blow for the theory of biological macroevolution? After all, the whole point of this theory is to explain all of biology. Can Darwinists accept the limits on neo-Darwinism imposed by the ID proponents?

Indeed, Darwin wrote in one of his letters that if his theory of natural selection required any miraculous additions, he would reject it as rubbish.

Intelligent design is not a "miraculous addition," because we can imagine that the intelligence that designed biological structures belongs to our universe. Theoretically, even an external intelligence would not need to use miracles to introduce design into nature. But for Darwin that would not be enough. Darwin boasted that his mechanism of variation and natural selection does not require any intelligent or purposeful causation. This is why he would reject the theory of intelligent design.

However, you should know that the British scholar also indicated how to falsify his own theory. He said that if it could be demonstrated that any complex organ existed, which could not possibly have been formed by numerous, successive, slight modifications, his theory would absolutely break down. Modern biochemistry has shown that there are many such organs. Indeed, every cell in such an "organ."

Darwin did not expect that because he did not know about, and so could not appreciate, the molecular level of life. He—along with other scholars of his time—assumed that if something is smaller, it must be simpler. Yet, in biological reality it is exactly the opposite. At the basic level of life, such as the cell, we find extreme complexity and sophistication that seems to be simplified at the level of whole, multicellular organs.

This means that in biology complexity precedes simplicity, and this contradicts widespread evolutionary assumptions. When we finally opened up the cell to investigation, we found devices that we knew about from daily experience because we already built and used them—electric motors, pumps, levers, gears, an advanced transportation system, the digital program in the DNA, etc. Even if you set aside the scientific arguments for ID, claiming that all of this came about without any purposeful agency is extremely naïve.

What novelty does ID theory introduce into the creation-evolution debate?

For almost a century the debate over evolution was presented as a conflict between creationism and Darwinism. No doubt, if we understand Darwinism as a mechanism which drives biological macroevolution, then

Darwinism and creationism are mutually exclusive. Either species were created directly by God, or they evolved from previous beings by some material process. These two ideas cannot be reconciled.

Note, however, that defining the controversy in this way may easily lead to a confusion between religion and science. And, indeed, this kind of confusion took place. Creationists spoke about "evidence for creation" and used the Bible as a source of natural knowledge. Given that creationism is usually understood as Young Earth Creationism, the controversy about the age of the universe, which is a scientific problem, has often been confused with the controversy about the origin of species, which is a theological problem.

Generally speaking, the problem with the debate was that science and religion—two complementary forms of knowledge—were often placed in opposition. The true conflict, however, is not between science and religion but between evolution and creation, that is, between two different concepts of the origin of species. On the evolutionists' side there is a tendency to supersede religion with science, while on the creationists' side, there is either biblical fideism or the reduction of religion to natural science (so-called creation science).

Only the emergence of the theory of intelligent design enabled scholars to set up the debate properly. Now, one scientific theory—Darwinism —is opposed with another scientific theory—intelligent design. In this way, the debate may be conducted at the level of natural science alone.

Doesn't this reduce the entire debate about evolution solely to the level of natural science?

This is a temptation for both theologians and scientists. The former can easily exempt themselves from undertaking challenging debates by saying: "Let's leave the scientific controversies to the scientists." The latter would gladly take over the competences of theology and entertain the false conviction that science can ultimately explain the origin of the universe and all its forms. Yet, science by its very nature has no power to explain the origin of species, and this is why the debate needs to be supplemented with philosophical and theological explanations.

In theology there are currently two competing concepts of the origin of species. One is theistic evolution (TE) another progressive creation (PC). I am leaving aside two "extreme" positions—Young Earth Creationism and atheistic evolution. Both ideas (TE and PC) are theological in character, so their supporters can pursue fruitful dialogue without falling into false alternatives.

The task for theologians is to debate which of these theological concepts better harmonizes with Holy Scripture and Tradition, and ultimately also with scientific data. I am a theologian, so I primarily deal with the theological level of the controversy about evolution. Nevertheless, I accept intelligent design, because this scientific theory is more compatible with the Christian concept of creation.

At the same time, I do not think that theistic evolution is an orthodox Christian idea. Therefore, the currently dominant duo of Darwinism in science and theistic evolution in theology should be replaced with another duo, namely, intelligent design in science and progressive creation in theology.

Most scientists refuse to call intelligent design a scientific theory. So, how can it be on the same cognitive plane as Darwinism?

When scientists wonder whether or not a given theory is scientific, they are acting not as scientists but rather as philosophers of science. Philosophers of science consider intelligent design a scientific theory more often than biologists do. Even physicists, mathematicians, and chemists are quicker to acknowledge the scientific status of ID.

Biologists who deny to intelligent design the status of science confuse science with naturalism; then they confuse naturalism, in turn, with materialism. I do not think that ID violates the principle of methodological naturalism, because I do not think that "intelligent causation" means the same as "supernatural causation." The question of whether or not ID theory violates methodological naturalism depends on how we understand the word "nature." If "nature" refers to matter alone, then ID conflicts with naturalism (granting that mind cannot be a material substance).

But under this assumption many theories commonly recognized as scientific would violate the principle of methodological naturalism. For example, all of mathematics makes sense only if we assume that there is some intersubjective ideal realm governed by unchangeable rules of reasoning which apply equally to everybody. Most people would agree that this realm does not exist materially. Nevertheless, we do not say that mathematics is supernatural, or that it violates the scientific principle of naturalism.

Similarly, the intellectual activity of man is something natural, despite the fact that the human mind is not material. Therefore, whether ID conflicts with methodological naturalism depends on how we understand nature.

Let's assume that ID theory violates the principle of naturalism and for this reason it cannot be called scientific. Instead, we must consider it a philosophical theory. Does this exclude the possibility of dialogue with Darwinism?

I would not cross swords about whether or not ID is naturalistic. Intelligent design contains both scientific and philosophical layers. But it is exactly the same with Darwinism. ID is a paradigm of doing science that includes some specific philosophical proposals, as well. Darwinism is also a paradigm in biology that includes some particular philosophical ideas. Therefore, both Darwinism and ID may be approached as either a scientific or a philosophical idea.

The juxtaposition of Darwinism and ID is justified and methodologically correct not because both theories are scientific, but because they are both on the same cognitive level and address similar questions. If someone does not consider ID to be science but philosophy, he must realize that Darwinism is also a philosophy. If one thinks that Darwinism is a strictly scientific theory, he must address the strictly scientific arguments against Darwinism developed by the proponents of intelligent design.

> *You said before that we can recognize whether or not a theory is scientific by checking what kind of question it addresses. You said that a theory that aims at explaining the origin of species cannot be scientific. How does this claim apply to intelligent design? Is it not a theory aimed at explaining the origin of species? If so, how could it be scientific?*

ID theory does not tell you *how* the design was introduced into biology. It only says that design *is present*; therefore, it must have been introduced (unless it exists eternally, which is hard to maintain knowing that the universe is not eternal). In other words, according to ID, there are elements of the universe such as irreducibly complex biochemical systems, genetic information, and other biological structures which are best explained by recourse to intelligent causality.

> *According to its critics, ID does not provide any material mechanism that would explain the presence of design in nature. So, how can it be a scientific theory without providing any mechanistic explanation?*

It is precisely this lack of a mechanism which makes intelligent design a scientific theory. It simply stops where the competence of science ends. If it attempted to fully explain the origin of species by explaining the nature of the intelligence or *how* intelligence introduced design into biology, it would not be a scientific theory anymore because it would enter philosophical and theological domains.

So, this theory not only remains within science, but also marks the boundaries of scientific explanations regarding the origin of species. Unfortunately, we cannot say the same about Darwinism. Darwin thought he explained *how* the apparent design was produced in biology. By proposing a mechanism Darwin reduced his explanation to a material process which by its very nature cannot create new designs. This is why his theory is reductive and thus cannot be true. Any theory that answers questions of origins cannot, by the very nature of such questions, be strictly scientific.

8. Darwin and the Death Camps

What is social Darwinism?

Social Darwinism is an attempt to apply the biological principles of Darwinian evolution to human social and cultural spheres. It stems from the conviction that as nature achieved incredible results in the form of different species, in the same way human culture and society will also achieve similarly amazing results if people adopt the same principles that govern nature.

Biological Darwinists believe that all living beings emerged through the struggle for life and the survival of the fittest. Drawing on this, social Darwinists believe that the struggle for life is an integral element of society. Its progress is propelled by the survival of the fittest.

How did it happen that Darwin's biological theory was applied to the social sphere?

In fact, it was the other way around—from a social theory to the biological one. In the early nineteenth century, Thomas Malthus published *An Essay on the Principle of Population.*[1] According to Malthus, human population grows geometrically whereas the resources necessary for life, such as food, grow arithmetically. The growth of human population will inevitably encounter limitation in the form of famine and disease. Thus, the idea that fertility leads to massive death was first proposed in the context of human society.

Darwin took this idea from Malthus and applied it to the plant and animal world. Later on, the "Darwinized" version of the Malthusian theory returned to philosophy and political practice as social Darwinism.

From the Christian perspective the common point essential to both theories is that they see struggle and death as key factors stimulating social and biological progress. From this perspective, the destruction of life, whether caused by natural or guided selection, is beneficial and serves a good cause. It may be an unwanted yet unavoidable condition for the growth of civilization.

How does social Darwinism influence our understanding of death? Can we say that death acquired a new meaning in social Darwinism?

Yes, absolutely. And this is the core of the mental revolution caused by Malthusianism and social Darwinism. In Christian culture death was considered evil—God neither made it nor does it please Him. Christian societies referred to victory over death accomplished by Christ through the Resurrection. Christians strive to save each person from both spiritual and physical death. This is why institutions such as hospitals, nurseries, and orphanages were invented and thrived in Christian Europe.

In contrast, social Darwinism puts death in a positive light because it works for the "improvement" of society. I believe that precisely the attitude towards biological death is what greatly distinguishes Christianity from Malthusianism and Darwinism. The same kind of thinking underlies the philosophy of some secular existentialists, such as Friedrich Nietzsche or Jean-Paul Sartre. This strain of Western thought culminated in Nazism, where we observe a fascination with death.

Today some Catholic intellectuals give credit to Charles Darwin, not only for being a Christian, but also for being a theologian who studied to become a religious minister. Malthus was also a theologian and even an Anglican minister. But crediting anyone just for being close to church institutions or studying theology is illusory. Obviously, not everyone who is nominally Christian follows Christian principles in his life or in his intellectual endeavors. In fact, Malthusian and Darwinian ideas provided ideological background for the worst aspect of Europe's history.

Was eugenics an attempt to further natural selection in society?

The founder of modern eugenics was a cousin of Charles Darwin named Francis Galton. The thinking of the first propagators of eugenics might be summarized in this way: "Natural selection serves to evolve better forms of animals, including man. Of course, we care about promoting a better humanity. Evolution endowed us with the faculty of reasoning and therefore we can guide natural selection. In order to foster evolution in society, we need to guide it, that is, we must allow only the best individuals to propagate."

Based on this principle, hundreds and thousands of people were mutilated in Germany and the USA in order to make their reproduction impossible. The decisions about who should reproduce and who should not were somewhat arbitrary. In the USA and Great Britain, the term "feeble-minded" was coined to refer to a mentally handicapped person. In reality, however, decisions about who should be included in that group were based on vague criteria. The British government considered "feeble-minded" anyone who was incapable of competing on equal terms with their peers or managing themselves and their affairs with "ordinary prudence."

You can see that this kind of definition opened the door for the abuse of power that actually took place. In the USA different local commissions deprived certain people of the "right to be normal" and coerced sterilization. In the Third Reich these practices flourished in the name of building a "new nation." Programs of forced euthanasia, abortion, and forced fertilization were implemented. Eugenic abortions were practiced in different countries. Altogether, eugenics affected many innocent people. Much of it remains an untold story.

Was Darwin a racist?

Charles Darwin worked with two close colleagues. One of them, Thomas Henry Huxley, was nicknamed "Darwin's Bulldog." Huxley held a high positions in British education and successfully introduced evolutionary theory into the textbooks. In fact, the theory never won the academic debate; rather, it was forcibly implemented by the authorities who governed education.

Darwin's other colleague was Ernst Haeckel who popularized his ideas in Germany. If we call Huxley Darwin's Bulldog, Haeckel should be called his Rottweiler. He was the first to present an advanced theory of human races. He explicitly claimed that the lowest of the human races sits closer to apes on the evolutionary ladder than to the most advanced humans. Darwin sent Haeckel his books with personal dedications and the two stayed in close touch throughout Darwin's career.

At some point, Darwin was worried that the theory of natural selection would be rejected in England. He was confident, however, that it would be accepted in Germany and thus survive thanks to Haeckel's engagement. In the *Descent of Man*,[2] Darwin wrote that in the near future, the civilized races would exterminate and supplant the savage peoples around the world. And indeed, an extermination did take place a few decades later, during World War II, when in the name of perfecting humanity certain human races considered inferior were murdered on a massive scale.

If we understand racism as the idea that different human races differ not only culturally but also, so to say, ontologically, i.e., they do not constitute one human species, then—yes, Darwin was a racist. This is implied by the very nature of his theory.

Can you say more about the connection between Darwin and Hitler?

First, I will point out one myth circulating among Catholics who are interested in Darwinism. It is sometimes claimed that Darwin's theory was abused by the Nazis and that Darwin himself cannot be blamed for all the horrible crimes of Nazism. Catholic evolutionists encourage us to clearly distinguish between the biological theory of Darwin and the philosophical idea of Darwinism. This thinking might be summarized as "Darwin—yes; Darwinism—no." By applying this distinction, it is easy to justify Darwin, because whatever evil came from his theory, Darwinism was at fault, not Darwin himself or his theory. In other words, they claim some ignoble ideologists used the biological theory of Darwin to implement their own wicked ideas into the social fabric and politics.

I think, however, that this line of defense of the Darwinian theory misses the point. It is like saying that the communism of Marx and Engels was essentially good, it was only Stalin or Mao who abused it. True, Marx did not murder millions; however, to say that there is no connection between Marx and Stalin is as untrue as to say that there is no connection between Darwin and Hitler. Communism led to the abuse of the power of the state, not because a few over-ambitious politicians misunderstood it, but because it was based on a wrong philosophy of society and human nature. It is not that communism provided those leaders with just an ideological background; it also shaped their thinking because ideas have consequences. These connections have been well documented by Prof. Richard Weikart and I recommend his books to those who want more historical evidence.[3]

Unfortunately, the connection between Darwin and Hitler is almost unknown in the popular consciousness. I think that Darwinism, similar to communism, is still an idea that really shapes our culture, especially the world of biological sciences. And that is why scientists who believe in the idea of biological macroevolution are not interested in revealing the inglorious pages of Darwin's theory. One can only hope that just as communism fell because it was based on a false anthropology, so Darwinism will fall because it is based on a false metaphysics.

*Is the distinction between Darwin's theory and Darwinism only
apparent? Does Darwin's theory necessarily lead to abuses of
human dignity?*

The distinction is not just apparent, because there is a difference between
a social philosophy and a biological theory. Note, though, that it is
ultimately irrelevant whether we can theoretically distinguish Darwin's
biological theory from Darwinism. What really matters is whether we can
maintain this distinction when we turn to social and scientific practice.
History inevitably proves we cannot.

So, what is the value of the theoretical distinction if the biological
theory itself shapes social attitudes and moral choices? Darwinism is—
has always been and will always be—used as an ideology or a paradigm
justifying and promoting certain views on society, philosophy, and even
religion. The distinction between "Darwin's theory" and "Darwinism,"
the way it is understood by Catholic evolutionists, implies that it is irrelevant what metaphysics you accept. What matters, instead, is your ethics.

In their thinking, Darwin's theory may represent an incorrect "metaphysics"—an incorrect theory of nature—but even the acceptance of
this incorrect "metaphysics" does not necessarily lead to the adoption of
mistaken ethics. This thinking is misleading because, as we have already
said, our ethics heavily depends on the metaphysics we adopt. We cannot
say what human rights are if we do not know who a human is. The dignity
of man depends on who or what a man is.

*It seems that today we can safely say that the abuses of social
Darwinism belong to the distant past. Nazism has been universally
condemned and we most likely are not endangered by a return of
forced sterilization. Are there any other dangers coming from social
Darwinism in our time?*

Even today there are philosophers and scientists who, in the name of
Darwin's scientific principles, want to decide about the life and death of
others. Leading proponents of population control like British Darwinist

Sir David Attenborough refer to human beings as "a plague on the earth" and demand a radical reduction in the human population. Proponents of "transhumanism" like Nick Bostrom, former Oxford University professor, and biologist Lee Silver at Princeton University are calling for a revival of eugenics to breed a superior human race using Darwinian principles. Extreme proponents of animal rights, like Peter Singer at Princeton University, argue that adult pigs are more valuable than human babies and justifies the killing of handicapped human newborns.

The fact that these people are not currently in power should not appease us, especially given that their postulates are not fading but are gaining popularity and support within some political circles.

Does the simian origin of man diminish our dignity?

If we accept that all organisms are related through biological generation, as occurs in the natural succession of individuals, it follows that every bee, snail, or ape is a cousin of man. And this is not a metaphor. A spider is your cousin completely literally—a distant cousin, but nevertheless a cousin. This is not a philosophical interpretation of Darwin's theory, but the very core of its claim that all species descended from one ancestor through natural generation.

But this raises a problem. If you kill a spider, should you be tried as if you killed a human being? If you avoid punishment, that means we don't treat all relatives equally. What criterion decides that harming a closer cousin, such as another human being, should be punished fundamentally differently from harming a more distant cousin, such as a spider? Our laws do not make a substantial differences between killing a relative and killing a stranger.

This thinking may seem absurd. But it is precisely this kind of thinking that underlies many contemporary cultural phenomena such as the radical animal rights movement, the green ideology, or the general tendency to obliterate the difference between man, animal, plant, and even machine. Therefore, the belief in the common descent of man and ape inevitably weakens the objective standards of moral judgment, regardless of whether or not anyone feels personally offended by this idea.

> *When interviewed by a Catholic magazine, the late Archbishop Józef Życiński asked in a rhetorical way, "Why should I feel insulted by this communion in genealogy with my brothers, who are lower animals?" And in response, he said that "we give into a certain megalomania, we would like to play a role so exceptional that no one in the universe is similar to us." According to the bishop, the simian origin of man does not diminish human dignity at all. What do you think about this?*

I think that what late Archbishop Życiński called "megalomania" is actually the classic view of Christians and Jews that is expressed in Psalm 8: "O Lord, you have made man little less than angels, crowned him with glory and honor. You have given him rule over the works of your hands, put all things at his feet."

Christianity infiltrated by Darwinism speaks about the community between humans and apes, whereas the authentic Christianity dares to speak about the community of humans with angels. Clearly, these two views of a human being are very different. Man, even from a purely biological point of view, is an absolutely phenomenal creature, transcending nature by his universal body that, unlike in other animals, is fitted to all kinds of physical activities that allow him to actualize his intellectual capacity. For example, human bipedalism is inexplicable within biological categories alone.

Man's resemblance to God is especially manifested in his intellect and in the rational soul. If you accept that God formed man in a special way without employing any secondary causes, then you will almost automatically attribute to man a great dignity. This is the work of God himself! Can something that resulted from the immediate Divine action be imperfect or defective? Does the work of God require any perfection by material processes? Similarly, if you believe in the separate creation of animals the great dignity of man does not detract anything from the dignity of the animal kingdom, because they, too, are incarnations of Divine ideas, designs that come directly from God.

In contrast, according to theistic evolution, man is a product of the struggle for life and the survival of the fittest. This means that there was no original perfection of man, but rather a gradual climbing up the ladder of evolution. Theistic evolutionists attribute the dignity of creatures to the community of atoms. The late Jesuit Fr. George Coyne, who was a proponent of the "Christian version" of Darwinism, said that "we all literally come from the same dust." By this he meant that all humans and everything else originates in interstellar dust from which—as the grand materialistic story has it—all the elements, the Earth, and all its beings were formed by the processes of stellar and organic evolution. The late Archbishop you quoted said also that "the same electrons that orbit in different objects, the same carbon understood as a chemical element, is found in the proteins of simple organisms as well as my own body."

This physical fact, according to him, should make us more open to the acceptance of universal common ancestry, and especially the simian origin of man. It is hard not to see how reductive this view is. It flips the order of creation upside down. The dignity of beings is deduced from motions in matter rather from than Divine wisdom. The metaphysics behind this view is a metaphysics deprived of natures and substances—individual beings are mere collections of atoms. If we follow this logic, there is no difference between man and a chunk of coal, because the same atoms of carbon build both. Ultimately, everything is the same and nothing has any special place in nature.

This boils down to absolute ontological relativism. Of course, we can keep asking, why should this detract anything from human dignity? But this is like asking, "Why is evil evil?" or "Why should we do good and avoid evil?" If someone does not accept first moral principles, it is difficult to find arguments that would convince him of anything else. Similarly, it is difficult to dialogue with someone who claims that the simian origin of man does not detract anything from his dignity. I think it takes a lot of effort to convince oneself to stop seeing the difference.

Do you see any specific dangers resulting from the acceptance of the
animal origin of man by Christian theologians and philosophers?

First of all, let's consider what the main threats are to human dignity in contemporary culture. I think it's safe to say that there are several such areas where the main struggle is being waged between contemporary Christianity and something that St. Pope John Paul II called the "culture of death."

First of all, there is the battle over the understanding of marriage and of human sexuality and the respect for human life. Note that Jesus already encountered people who promoted errors about marriage. The Pharisees asked him whether it was lawful to divorce your wife. They referred to the law that Moses gave them. In his response Jesus referred to the Genesis account of creation. He said that in the beginning it was not like that. In the beginning God created man and woman to be one. For centuries, the Church followed the path indicated by Christ, that is, the Church referred to the historical understanding of the creation account. This was still the case in the nineteenth century, when divorce began to be legalized in Christian societies. Catholic theologians argued against this legal destruction of marriage by referring to the Book of Genesis.

The argument from Genesis is not about some religious beliefs of one social group. If the account of creation is true, then it is universal and applies to every human being. All people share the same nature. This classical Christian perspective is beautifully depicted in the encyclical *Arcanum divinae sapientiae*, which I mentioned before. Leo XIII referred to the account of creation precisely when he defended the institution of marriage. And how is it today? Nobody quotes Genesis anymore, because we "know" that these are just metaphors, reminiscences of ancient mythologies. If the message of Genesis is not true, meaning it does not correspond to any historical reality, then of course it cannot constitute any argument in the debate.

Does this mean that the Church deprived herself of an important argument in defense of marriage by giving up the historical understanding of the creation account? How should we pursue the public debate on marriage?

Obviously, there are many arguments in favor of the traditional understanding of marriage that are offered by social sciences, philosophical ethics, and even medicine. Nevertheless, the ultimate justification for human dignity in the public debate should come from the historical understanding of the human origin as depicted in Genesis. This story is common to all people because all people descended from Adam and Eve. All humans have them as parents, including unbelievers, as well as those who believe in Darwin's theory.

Today the problem is that the dominant tendency among Catholic theologians is to justify human dignity by emphasizing the rational soul in a human being or the human capacity of having a relationship with God. As true as these observations are, they are also reductive, because they shift the evidence for human dignity from what is visible and historical toward the invisible and spiritual alone. This is an easy way to avoid the hard topic of evolution and human origins. But there is a cost to this "easiness," consisting of the fact that Catholic apologetics becomes ineffective and irrelevant, because it adopts the same naturalistic worldview on which the secular ideologies are founded.

How does the adoption of the evolutionary paradigm make Catholic apologetics irrelevant?

Most theologians say that the body could have descended from an animal ancestor, but it is much more important that man has an immortal soul which makes him the master of creation and safeguards his special dignity among other creatures. The problem is that these kinds of arguments meet with disregard or indifference from the cultural opponents of Christianity. Their response is something like this: "If your faith tells you only this much, why are you bothering us in the first place? If you want, you can believe in your spirits, souls, or any other superstitions. The point is that you do not postulate anything about the visible and historical world. This is our territory. Let there be any kind of soul you like, we do not care, because science tells us nothing about it. We know that Darwin explained the origin of man."

And, unfortunately, if Christianity were all about just the invisible realm, then those unbelievers would be right. But that is not the case. Christianity makes strong statements about the visible realm, as well. It speaks about historical events that happened in the natural world, but exceeded the order of nature. It also speaks about the ultimate destination of the universe and man, which also exceeds the order of nature. This is why saying that human dignity is justified solely by the human soul is unsatisfactory. It does not do justice to a truly Christian anthropology.

St. Paul asked: "Do you not know that your body is a temple of the Holy Spirit?" (1 Cor. 6:19). This fact about the human body by itself provides a dignity that makes the human being incomparable with any other living organism. The renunciation of postulates about the visible sphere is precisely what makes contemporary apologetics weak and fruitless. Arguments "from the soul" fall on deaf ears. They are convincing only for those who are already convinced.

Atheists and agnostics think in materialistic categories. To win them for Christ, first, we in the Church need to return to the Christian understanding of man regarding both his physicality and spirituality. The return to a true understanding of human origins as revealed in the Bible would obviously create enormous controversy. But the more neglected the garden, the more effort it takes to weed it. Avoiding controversy cannot be a guiding principle for Christian apologetics.

In what way can the historical understanding of human origins as described in Genesis become an argument in contemporary moral and cultural debates?

Let's focus on the most obvious problem. It is said that two men may love each other in the same way as a woman and a man; therefore, they should have the right to an equal legal status of marriage. You may present many arguments to explain why this reasoning is false. The premise is not true and the conclusion does not follow. But the ultimate answer comes from our understanding of how man emerged in the first place. If the human body is a product of evolution, then human sex is something acquired on the long road of chance modifications. Initially, there was no sexual reproduction, and then, over an immensity of time and due to slight, organic modifications, two distinct sexes appeared.

Now, we have taken over the evolutionary process and we can shape our sexuality according to our whims—which means that sex is reducible to "gender." Sexual "identity" is a matter of culture rather than nature. You are a man because this is how you were brought up. You are a woman, because when you were a child, you were given dolls to play with rather than action figures.

This leads to a number of problems, including the definition of the social roles of man and woman, gender dysphoria, and the struggle for so-called women's rights (radical feminism). Then, it radiates to all spheres of culture, including even the language of theology (God as a woman). In more extreme cases, it leads to so-called "transitions," i.e., an attempt to change biological sex through chemical and surgical interventions in the body. The roots of all of it can be traced back to evolutionary thinking about biology and human origins.

Now, if you believe that man began to exist through the immediate formation by God, this implies that sex is something natural and unchangeable. Right from the beginning, humans emerged as a man and a woman—two complementary forms of humanity. Nature is established by God in the act of creation and this reality is unchangeable. Culture is what is changeable, because people create it, but human sex belongs to nature rather than culture; therefore, it cannot be changed.

Moreover, according to Genesis, the first man was missing something until the woman appeared, which means they supplement each other in marital community. In the beginning, they constituted the happiest couple ever. Each marriage, in a way, is an attempt to rebuild that original happiness. This is why only a community of one man and one woman can be a true marriage. All of this is there in Genesis, which is why the Church needs to return to the belief in creation.

9. Why Is Teaching Evolution So Controversial?

Why do propagators of Darwinian evolution fail to acknowledge the serious arguments against their theory?

This is a complex problem and there are different reasons, depending on who exactly we are talking about. Imagine that you are an atheist. You do not recognize any higher reality, let alone the creation of all things from nothing. If someone asks you, "Where did the world come from?," you answer that it did not come from anywhere. It simply exists from eternity. This answer is problematic today, because all scientific data point to a temporal beginning of the universe. As an atheist, you would not agree that the Big Bang was the beginning of the universe, but you do not have a better theory. The idea of cyclic "big bangs" and "big squashes" is a view that is not supported by any data. Modern atheism is based on faith and imagination rather than any facts.

Let's say, however, that you have dealt with this problem by adding a bit of agnosticism to your originally pure atheism. Now, you say that you do not know where the world came from, but once we have the universe as constituted by matter and the laws of nature, then everything follows naturally from there. Cosmic, chemical, biological, and social evolution explain the origin of everything—from bacteria, through all species, to man, his culture, and his laws. Your worldview is basically coherent.

Then, all of a sudden, someone comes along and says that a rather important part of this evolutionary story is questionable. All biological forms, and even more so life itself, could not have come into being through purely material processes, without any involvement of intelligence. Such a claim makes a breach in the atheistic worldview. For if intelligence is necessary to produce biological forms, it means that it existed before man came into being. That turns the evolutionary story upside down. Being an atheist, you cannot agree with that and modern atheists know this well.

For example, Richard Dawkins claims that atheism was logically possible before Darwin, but it was only Darwin who made it possible to be an intellectually fulfilled atheist. Notice, though, how the line of reasoning is reversed in atheism. The initial question was: How do we explain the origin of species? We were looking for the best theory. The atheist's starting point is the assumption that the answer must be purely naturalistic. And since Darwin's theory is the only widely accepted naturalistic theory, now their goal is to defend that theory rather than to find the best explanation. An atheist must defend Darwinism regardless of whatever arguments there may be in favor of or against the theory.

This is precisely what we are dealing with in popular culture today. There are many arguments advanced to support Darwin's theory, but I do not think that any of them are decisive or even convincing. In other words, they receive far too much credit. At the same time, there are many strong arguments against the theory, though you will not find them in textbooks or popular media coverage.

Father, you spoke about atheists. But very often also believing Catholics, theologians, philosophers of nature, and faculty members at ecclesiastical institutions defend Darwin's theory. Are they also deaf to contrary arguments?

I think the situation of Catholics who believe in Darwinism is more complex. Since the nineteenth century, theologians have been accused of not practicing science, but rather of fostering fairy tales. First positivism, then logicism, Marxism, and scientism became *de rigueur* for theologians. In Poland, mainly under the influence of Marxist ideology, theologians have developed something I would call a "complex of being unscientific." For years, they were told that theology is not science; therefore, it should be expelled from universities. There was no place for theology in academic journals or textbooks, let alone research grants.

Of course, communists had their own reasons for fighting theology and removing it from the academy and public life. But apart from the political dimension, it was a dispute about the status of theology in general. The response of professional theologians was typically to prove at all cost that theology is scientific. In doing so, they forgot that the word "science" itself has different meanings. It is obvious that theology is not science in the modern sense of the word, which means natural science. Theology derives its premises not only from the natural world, but above all from supernatural revelation.

This does not mean that it is subjective or that it does not convey any knowledge. In fact, its claims are more certain than those of the natural sciences precisely because they rely on the authority of God. Therefore, the fact that theology is not science in the same sense as physics, chemistry, or biology does not detract anything from its value as a cognitive discipline.

However, theologians have generally taken the path of proving that theology is science and therefore should have the same rights and privileges as all other scientific disciplines. Hence, the hypersensitivity of theologians to accusations of being unscientific. I think you realize how easily the "complex of being unscientific" can be used by nonbelievers to play on the issue of evolution.

So, you are saying that if a theologian challenges any element of Darwin's theory, he will be accused of "being unscientific." This must serve to silence him.

Exactly. And let him not even think that there might be some theological concept of the origin of species! If he claims there is one, he is not just unscientific, he is altogether "anti-science," which means that what he says is meaningless, simply stupid.

Unfortunately, the "complex of being unscientific" also works the other way around. Want to be scientific? All you have to do is embrace the "consensus" of the scientific community and mindlessly repeat that God could have used evolution. It doesn't matter how you reconcile this with Genesis or how it relates to sound philosophy, much less whether evolution even occurred. The important thing is that you are "open to science."

Does the scientific community take theistic evolutionists seriously?

I think that in scientific circles, theistic evolutionists play a role akin to comics who entertain scientists with tales of how the entire world is filled with divine presence. This claim, true in itself, does not tell the whole truth about the Christian understanding of origins. In discussions about evolution, theistic evolution plays the role of a quasi-religion—and not an entirely serious religion, either, one easily reconciled with pure naturalism. Staunch Darwinists take all this with a grain of salt. They know that God is not needed for anything, but why would they dismiss such a playful religion, especially when it does not change anything in their worldview?

Theistic evolution is uncontroversial in scientific circles because it is Christianity devoid of what is Christian. Some of the Darwinists in these social circles aren't pure atheists, so the company of a theistic evolutionist—especially, if he's a clergyman—helps them feel more comfortable. Not only do they cling to a naturalistic worldview, but they also find a priest telling them there's no problem with doing so. It's a bit like inviting clergy to participate in left-wing journalism. Does anyone there care about preaching the Gospel or explaining the Church's teachings? No. The point is that every show needs an entertainer, a character who comes, dances to the music while everyone laughs, and then walks away. That's the role theistic evolutionists play in the circles of scientific materialists.

Are there any other reasons for theologians to defend evolutionism?

Generational continuity plays a crucial role in the Church. Evolutionism dominated the departments of Catholic theology around the mid-twentieth century. This means that the "Treatise on Creation" (De *Deo creante*) has not been taught in the Church for over seventy years. It has been replaced by theistic evolutionism, which states that it doesn't matter "how" God created, what matters is "that" God created. Some of the more conservative theologians keep silent over the question of the origin of species because they either do not know what to think or have an intuition that macroevolution is incorrect but feel that challenging it costs too much.

Sometimes, instead of answering questions concerning origins, various models of science and religion are presented. In each of these situations, what was taught for centuries is no longer taught at all. As a result, we are now dealing with the third or fourth generation of professors of theology who do not know the theology of creation. So, how are priests, catechists, and lay faithful supposed to know it?

For example, no one with even a basic understanding of the Christian view of creation would argue that God could have used a secondary cause, such as evolution, in creating the world. Unfortunately, education abhors a vacuum. If sound theology isn't taught, its place is taken by the most bizarre propositions, such as theistic evolution, which is bad science and even worse theology. It is difficult to say who is to blame for this state of affairs, but one thing we can be certain of is that the matter requires clarification within the Church herself.

How is it possible that Christian scientists defend Darwinian ideas? There are even instances when they become more passionate about defending Darwin's theory than any biologist.

Once someone immerses themselves in evolutionary thinking it's difficult to shake it off. If they don't try to disentangle themselves from this way of thinking, after some time they begin to see everything in evolutionary categories. If, on top of that, they work in a field where they are constantly told that "evolution is a fact," they eventually begin to believe it. Evolutionary thinking becomes their identity. It doesn't matter whether everything actually arose through evolution; what matters is that they believe it was so.

And because they strongly believe it, they start treating it as "fact." A strong desire for something to be true causes a person to believe it is true, regardless of empirical confirmation or a lack thereof. Once their mind is completely captured by the belief in evolution, they put a considerable amount of effort into convincing and deceiving others. How could what I believe so strongly not be true?

And if someone publishes a few papers explaining how evolution can do this or that, it is hard to back off. The further they go, the harder it is to turn around. So even when those rare moments of honesty come, they think it is inappropriate to publicly deny what they just defended so fervently. I think that if scientists who actually follow the debate on Darwinism continue to believe in Darwinism, it is more likely due to some kind of mental inertia or some other non-scientific reasons.

And how do theologians react to the critique of evolution?

For theologians, evolution is a matter of worldview. Such topics typically provoke more emotion. And that is why, when someone questions evolution in the presence of theistic evolutionists, they feel personally attacked. Typically, they start aggressive apologetics for Darwinism. And at some point, you wonder what is going on. If Darwinism is truly just a biological theory, why does presenting rational arguments against it infuriate theologians? You begin to suspect that the stakes are much higher. And indeed they are.

Theologians defending evolution are not really defending a particular theory, but an entire theological vision on which they have based their views and perhaps even their personal lives. According to their view, God, as they say, "does not intervene" in the world. Of course, a world in which God does not intervene is not very different from a world in which God does not exist. But it is very different from a world in which God does "intervene."

If God limited himself to creating just the Big Bang, then everything else is produced according to the laws of matter. The universe is predictable and—in a way—subdued to our minds, because we know the laws of nature. But if God acts beyond that, then we must acknowledge that we may be surprised, that we are not absolute masters of the world, because there is someone higher up who is the absolute Ruler of it all.

If the Ruler truly reigns, and does not just observe from behind the glass window of his palace, then we must reckon with possible "interference" in my life, as well. Freedom is no longer arbitrary. God can exact punishment for the evil I have committed; He can change the course of my life; He can put new challenges in my way. The failure to recognize God's true rule over the world constitutes a deeper reason for rejecting the original Christian belief in creation and replacing it with a belief in evolution.

Can we count on some kind of general "conversion" of theologians?

I don't think such an expectation is realistic. A certain generation simply needs to change. Paradoxically, it is easier to discuss evolution with scientists. They are used to the fact that scientific explanations change with better theories replacing worse ones. Theologians, on the other hand, are less inclined to change their stance because they know that in theology, once something is true, it is true forever.

Moreover, if someone has already published a book or articles arguing that biological macroevolution poses no problem for Christianity, it's difficult to retract it. Rare are people of St. Augustine's integrity who, towards the end of his life, was able to write an entire book about his own errors.

The late Archbishop Życiński argued that denying Darwinism leads to ridiculing the Church and Christianity. Can theologians therefore oppose science?

If they opposed science, that indeed would be a problem. But proponents of intelligent design do not oppose science; rather, they oppose a particular theory, proposed as scientific, called Darwinism. In turn, theologians who oppose theistic evolution also do not oppose science, but rather a certain theological idea, namely, an evolutionary reinterpretation of the classical theology of creation.

If someone believes that the guiding principle of Christianity should be avoiding ridicule, they should not speak out on any controversial issue. I do not know what Christianity would look like if Jesus had followed this attitude when he stood before the Sanhedrin. Unfortunately, there are also scholars who have left the Catholic Church precisely because of the support that the theory of evolution enjoys within it. Isn't this a terrible counter-witness for which we will need to apologize one day?

What future direction do you foresee for this debate in the Church?

It is not a secret that some members of the Church hierarchy are aware of the problem. They see that Catholic evolutionists like George Coyne did not espouse orthodox views. The hierarchs know that "something is wrong," but they cannot precisely define what.

This is evident, for example, in statements by Cardinal Christoph Schönborn, who was still the most precise among the hierarchs speaking on this matter. Benedict XVI's teaching could be summarized roughly as follows: "Evolution has certainly overstepped its bounds becoming a philosophy attempting to explain the entirety of reality." Unfortunately, the Pope neither indicated where these bounds were crossed nor made any attempt to rectify the situation. Pope Francis did not bring any novelty to this issue and it seems like the Church in her official statements is locked in the position presented by Pius XII in 1950.

In the world of science, however, there is no stagnation. Anthony Flew, Britain's leading atheist, used to be invited to various symposia throughout his career to explain how the idea of God is useless. Shortly before his death in 2010, he changed his views. He stated, among other things, that the argument from intelligent design is much stronger today than when he first encountered it. He admitted that new discoveries strongly supported it. This argument led him to the acceptance of the existence of God.

In 2012, another atheist, Thomas Nagel, published a book entitled *Mind and Cosmos: Why the Materialistic Neo-Darwinian Conception of Nature Is Almost Certainly False*.[1] The title speaks for itself. Nagel, who did not undergo any religious conversion, suggests that Darwinists must seek a new theory, because within the framework of pure Darwinism they cannot answer the arguments from intelligent design.

If atheists themselves are losing faith in Darwinism, something interesting is happening. Of course, these phenomena should not go unnoticed in the Church. I fear, however, that we will also witness a sad situation when many atheists abandon Darwinism, yet Catholic natural philosophers will remain on the battlefield "for the only true idea." It is likely that some Darwinists will abandon the belief that species were shaped by the blind forces of nature and adopt instead an old view that demonic forces had something to do with it. After all, a demon is also some kind of intelligence. It is not impossible that some Catholic evolutionists will also follow this path.

By appealing to the demon, they do not have to accept the despised "divine interventions." But such a shift from "pure chance" to an "evil spirit" would make the hierarchy realize that the debate on evolution is not and has never been a scientific debate, but a religious one. It is a debate about God's role in the formation of the universe. The view that plant and animal species were shaped by the devil is not new. It appeared in the Middle Ages and was condemned in the fourteenth century. I think that ultimately theistic evolution will be condemned by the Church.

We talked about the debate in the Church. Let's move on for a moment to that second debate, which is taking place in the public square, education, and the media. Some of the most heated verbal exchanges among internet users are precisely over the question of human origins and evolution. Why does this issue evoke such strong emotions?

Because, as I said, it's not a scientific issue. If the dispute were between two scientific theories, it wouldn't have spread beyond a narrow circle of specialists. Everyone has their own opinion on evolution, because it's not about a particular biological concept. One's attitude toward evolution determines a whole range of intellectual preferences. Are you a believer? If so, which God do you believe in? How do you understand the Holy Scriptures? What role does science play in explaining reality? What can nature or pure chance do? Can science be wrong? And so on. By taking a stance on biological macroevolution you simultaneously define your preferences in many other moral, ideological, philosophical, and religious debates.

Should biology textbooks teach creationism?

We are talking about public education, meaning it is funded by all taxpayers. Among them are those who strongly believe in evolution, those who have no opinion, and those who don't believe at all. But if we agree that our science classes convey any objective knowledge, then textbooks must be standardized, meaning the same for everyone.

As I already said, evolution (understood as macroevolution) is irrelevant to the science of biology. It's a paradigm through which students are forced to see all biological phenomena. If evolution were removed from textbooks, they would still be biology textbooks, and the solid science contained in them would lose none of its scientific value. But if it needs to stay there because it cannot be removed, arguments should be presented in support of it, as well as against it. There are several so-called "evidences" for evolution that are reproduced in most textbooks. But all of them are based on false data, or a misinterpretation of data, or a failure to prove what they are supposed to prove. For a detailed discussion of these "evidences" I refer you to Jonathan Wells's book, *Icons of Evolution.*[2] Wells analyzed American textbooks but the same examples appear in the Polish ones and I presume in many other countries. Therefore, taxpayer-funded textbooks used in public schools promote false knowledge.

One might ask, Where are all the committees that review the quality of these materials and approve their use? The demand for "balanced education," i.e., the presentation of both evidence for and against evolution, is entirely justified. I'm not talking about introducing creationism. Let's simply allow science to speak for herself. Students have the right to hear the arguments from both sides.

10. Was Eve Created from Adam's Rib?

The origin of man from the dust of the earth is already difficult to accept, but it seems that accepting the creation of Eve from Adam's rib goes beyond reason. Most theologians consider this a childishly naive view of the Holy Scripture. So, how should we understand the biblical account of the creation of the first woman?

Many theologians believe that the creation of woman from Adam's side is a metaphor. They say that the Holy Scripture cannot be taken literally. Sometimes, they ridicule such an interpretation, and if they do not ridicule it themselves, they are very worried that a non-believer who hears it would laugh at such a naive faith on the part of Christians. These theologians argue that it is not permissible to say that Eve was created from Adam's rib, because this leads to ridiculing the faith. They think they defend Christianity, but they do it in such a way that they directly deny Holy Scripture. Moreover, many of them believe in different Catholic dogmas, such as that a wafer (bread) is turned into the most holy Body of Christ every day on altars around the globe. So, why can't they accept that there was a one-time event when dust of the earth was turned into the human body? Frankly, I do not understand why the belief in the creation of our first parents is so difficult for people who believe (or at least claim they believe) in things that by themselves are way stranger and more incredible than the creation of Eve from Adam's rib.

Some claim that the biblical account should be understood more deeply somehow, not so physically, not so simplistically. But when we ask them directly how the first woman began to exist, we will never hear a direct answer. Some, for example, propose that Eve was an identical twin of Adam in the womb of the last hominid. Others would claim that they evolved separately, that is, just as God gave a rational soul to a male hominid, so He gave a rational soul to a female hominid. But I do not think that these concepts are more convincing than the biblical account. They encounter all the problems that we discussed when talking about the creation of man.

Additionally, when we abandon the biblical history of the creation of woman, some important theological content is lost. Therefore, those who abandon the Holy Scripture do not offer anything better in return. It is only surprising that there are people in the Church who are inclined to accept the most bizarre theories as true, just not the events recounted in biblical history.

And how does tradition understand the creation of Eve? Literally?

Let's start with Scripture itself. When I spoke about the creation of man, I referred to the analogy of faith—explaining one truth of faith with another. On the basis of analogy, one can also explain some biblical texts in the light of others. In the First Letter to the Corinthians (11:8), St. Paul writes: "Man did not come from woman, but woman from man." And further: "As woman came from man, so man is born through woman."

It is obvious that two thousand years ago these passages were not understood metaphorically. However improbable it may seem, a man does come out of a woman's womb when he is born, so he comes from her literally. And in the same way, the first woman came from man. St. Paul does not repeat the Book of Genesis uncritically, but summarizes its content in order to justify his teaching on covering the head during prayer. Therefore, St. Paul, like the entire Jewish and later Christian tradition, understood the creation of a woman from the side of man in a literal way.

What did later theologians say about this?

One could quote endlessly the Fathers and Doctors of the Church to support the thesis that they understood the creation of Eve from Adam's side quite literally. For example, St. Cyril of Alexandria argued to the Jews for the virginal conception of Jesus: "Whence, then, was Adam made? Did not God take dust from the earth, and fashion this wonderful frame? Is then clay changed into an eye, and cannot a virgin bear a son? Does that which for men is more impossible take place, and is that which is possible never to occur?" And further: "Of whom in the beginning was Eve begotten? What mother conceived her, motherless as she was? But the Scripture says that she was born out of Adam's side. Is Eve then born out of a man's side without a mother, and is a child not to be born without a father, of a virgin's womb?"

Note that if Cyril saw nothing miraculous in the creation of Eve, his argument for the miraculous birth of Jesus would make no sense. St. Cyril emphasizes that just as Adam was created from the dust of the earth, so Eve was made from Adam's rib. He is speaking here of the "matter" from which the human body was formed. So, this is no metaphor.

Similarly St. Ambrose writes: "It is not without reason that woman does not come from the same earth from which Adam was formed, but from the rib of the same Adam. [The reason is] that we may know that the nature in the body of man and woman is one, and that the source of human nature is only one." It is obvious that Ambrose too understands the formation of woman from a rib in a literal way. He contrasts the earth and the rib as two different materials. The first man was formed from the former, the first woman from the latter.

Another Doctor of the Church, St. Jerome saw the problem of the human origin in terms of faith and unbelief: "Does anyone believe in God the Creator? He cannot believe unless he first believes that to be true which is written of His Saints: that Adam was fashioned by God, that Eve was made out of the rib and from his side, that Henoch was translated ... He who does not believe all these things and the others that have been written about the Saints, will not succeed in believing in the God of the

Saints." Thus, Jerome treats the belief that Eve was formed from Adam's rib and side as an essential and integral element of Christianity. I do not know on what basis modern theologians argue that this is not an essential truth of faith.

On the other hand, St. Augustine, who is often cited by Catholic evolutionists, wrote, among other things: "God created only one man, not, of course, to be lonely, deprived of all company, but so that in this way the unity of society and the bond of harmony might be more effectively entrusted to him. In this way, people were bound together not only by similarity of nature, but also by family ties. And in fact, God did not create the woman, who was to be given to him as a wife, in the same way as he created man, but created her from man, so that the whole human race came from one man."

Augustine emphasizes that both man and woman were created, not born, but woman differently from man because she was created from him. In addition, Augustine gives a justification for why this happened in such a way. Namely, so that family ties would be stronger, and so that everyone would feel like a member of one human family. Notice, then, that the Tradition not only understands Eve's origin literally as being formed from Adam's side, but also confirms that this truth has great anthropological significance.

What anthropological truth does the formation of woman from Adam's rib express?

If you think carefully, you'll conclude that there is no other way of creating woman that would simultaneously ensure two things: the unity of the human race and the equal dignity of both sexes. Any other way of creating woman either diminishes the dignity of one sex (or both), or makes sex merely an apparent phenomenon leading to sexual "indifferentism," or disconnects woman from man, ruining the unity of the human race. I think that the concept of Eve being a twin of Adam proposed by some theistic evolutionists makes gender something only apparent and it does not retain the descent of Eve from Adam. The other concept—hominization of two different hominids—disrupts the unity of the human race and does not account for man being the origin of woman, either. Besides, all these concepts are essentially an offence to human dignity. If we were pagans, we might believe that man arose from the excrement of some gods, or as a result of copulation between heroes and goddesses, or something like that. However, when we take the mythologies and views propagated by theistic evolutionists on one side and the Christian biblical faith on the other, Christian anthropogenesis is simply more beautiful and appropriate for humans. Could there be any nobler origin of man than through the direct action of God? I can't even imagine such a thing.

The formation of Adam from dust is also of immense significance. God could have created man without the use of dust, simply by bringing his body out of nothingness. But then man would lose this fundamental connection with the material world. He would feel like some alien from beyond the world who happens to walk on this planet. Therefore, not only the direct action of God, but also the fact that God uses material in the form of dust, has anthropological significance. I think this is what John Chrysostom wanted to express when he wrote: "God formed man by taking dust from the earth. See also the honor paid to you in this. He took nothing else from the earth but dust, the smallest thing on earth that was. And at his command, he changed this dust of the earth into the nature of flesh. In the same way that he created the very substance of the earth when it did not yet exist, so now, by his own will, he changes the dust of the earth into flesh."

Chrysostom speaks of two stages in the formation of the human body. In the beginning, God created the substance of the earth when it did not yet exist. Here, he speaks of the creation of matter from nothing. But in creating man, God takes this very earth—"nothing else," as Chrysostom emphasizes—and changes it by his own will into flesh. It is obvious that this Father of the Church also treats the creation of man from the dust of the earth quite literally.

Did God take Adam's rib to form Eve, or perhaps some other fragment of his side?

That's a good question. And I think that if Christians weren't misled by modern mythologies, i.e., all these stories about hominids, we would have interesting discussions on this topic. St. Thomas Aquinas, for example, believed that God created Adam initially with an extra rib, which he later removed from his body to form a woman. This is not a dogma, and I don't think it should become a dogma. Nevertheless, Aquinas had his reasons for this assertion.

We can therefore consider four options. (1) According to the first view, God takes Adam's rib and restores it with flesh alone. Even though this option is consistent with Scripture, it would make Adam crippled for the rest of his life, which is not congruent with the perfection of divine creation. (2) The second option is that God takes matter from Adam's side—not the rib, but the flesh itself, without the bone—and as Scripture says, fills its place with flesh, that is, miraculously restores the missing tissue to Adam. This interpretation is possible in light of the Hebrew text and the Septuagint, but not according to the Vulgate, which uses the word "rib" (*costa*). (3) On the third view, God takes one of Adam's ribs and covers its place with flesh along with a new rib. In this case, the word "flesh" would have to mean both tissue and the rib. (4) The last option is that of Aquinas, who says that God created man with an extra rib, then removed that rib to form woman and filled the space left by the missing rib with flesh alone. It seems that this position is the one most consonant with the text, at least in its Latin version, and it does not leave Adam missing anything.

Assuming that Adam was first created with an extra rib, does it somehow impact our understanding of the creation of man?

If we assume that he was created with an extra rib, the truth that woman was immediately in God's mind is clearly highlighted. One could say that a man with an extra rib was somehow defective, but at the same time, in some sense, he carried a woman within himself. Only the creation of woman made man fully functional and, so to speak, normal. Therefore, the "extra rib" option points more to the necessity of Eve's creation.

I would venture to say that women and men with a more egalitarian attitude would be more inclined to this solution. However, the creation of woman from a rib that God later restored to Adam seems to emphasize more strongly the moment to which St. Paul refers, namely, a certain dependence of woman on man. Nevertheless, in both cases, we are dealing with the formation of Eve from Adam's rib. This, in turn, highlights the unity of the human race, which all comes from one man, as well as the call for mutual love between man and woman.

But doesn't the creation of Eve from Adam's rib diminish the dignity of women?

Equal dignity of the sexes does not imply their identity. Feminist activists would certainly prefer women to be hominids "enlightened" independently of the male hominid. In the most "creationist" scenario, they would probably want women to arise from dust, like Adam, or even better, without any prior matter. They would then constitute a "class in themselves," without competition, without dependence, and, unfortunately, without any ties. But this is the price of liberation at any cost. And I think, unfortunately, this scenario is also realized in the personal lives of many feminists. They do not want to be from the rib of one man, and therefore they cannot build a relationship with one man. They never belong to anyone, and therefore cannot be a gift or a specially chosen good.

But let's return to the question. Since equal dignity does not imply identity, the relationship established between woman and man at the moment of creation fundamentally benefits both of them. In other words, some dependance of woman on man, established by creating her from Adam's rib, is natural and, as such, beneficial to woman.

Furthermore, note that God took a rib, a tiny fragment of matter compared to the entire body. And this is significant. Adam was not deprived of anything essential: head, lungs, or heart, and thus the creation of woman does not diminish Adam's dignity in any way. At the same time, God has to add on much more of the woman's body than he took from man. The woman's body is therefore developed mostly from nothing, making her a work of God's power. Woman is not merely a "revamped Adam," but an original creation whose body reflects the Creator's original design.

Since we're going into such details, where did the matter come from to form Eve's body? Does theology have anything to say about that?

St. Thomas Aquinas compares this to Jesus' multiplication of the loaves. He considers two possibilities: either the matter of Adam's rib was rarefied, or God added new matter. Thomas rules out the first option because a woman's body does not differ greatly in its density from the body of man, at least not to the proportion of a rib to an entire body. Therefore, God added new matter, though not by creating it out of nothing, but by converting some original matter of Adam's rib.

St. Bonaventure understands it similarly to Aquinas when he says: "Someone might doubt that woman was created from a rib by the same miracle as the multiplication of the five loaves and others. It must clearly be said that the formation of woman was miraculous. ... Woman was created miraculously from a rib not because this happened against the ordinary course of nature or against nature, but because she was made by a divine power acting above nature. ... The reason why God formed her in this particular way is that the manner of formation was above the power of nature. And so that man in both sexes would turn to God directly and love Him with his whole being, knowing that he was made directly by Him."

Regardless of the metaphysical details that St. Thomas tried to unravel, Bonaventure's statement points to what is essential. First, woman came into being by a miracle, and therefore not by the operation of natural laws. Second, she came into being from Adam's rib, and therefore not from some hominid. And thirdly, woman, like man, was created directly by God, that is, not by virtue of secondary causes such as evolution. According to Bonaventure, the direct creation of our first parents is a sufficient reason to "love God with our whole being."

I once encountered an opinion that we are all created directly by God. How should this be understood in the context of St. Bonaventure's and St. Thomas Aquinas's statements?

The claim that we are all created directly by God is imprecise, as it suggests that each of us was created directly by God. However, this is only true in terms of our species. That is, all of us as a human species were created by God. However, in terms of each individual, only our first parents were created directly by God. All other humans are born from parents and therefore are not created. At the moment of conception, we receive the soul from God, but here God does not create any new species, but only individual souls belonging to the species of human souls. Therefore, the human individual, in terms of body, is not created directly by God; he is born from his parents.

I think this difference is well captured, for example by John Damascene, who writes: "The earliest formation [of man] is called creation and not generation. For creation is the original formation at God's hands, while generation is the succession from each other made necessary by the sentence of death imposed on us on account of the transgression."

Nine centuries later, the Jesuit theologian Francisco Suárez formulates it even more precisely: "When a man is formed by way of natural generation, the organization of the body and its orderly arrangement are the result of the natural power at work in generation. But, natural propagation apart, such a result cannot come from any created agency whatever, whether elemental, or celestial, or from all of them together. In the first man, obviously, the organization of the body was not produced by natural generation; its author, therefore, could be none but God alone."

Both statements confirm that Church tradition clearly distinguished between generation and creation. Only the first parents were created, whereas all subsequent humans are born. And I don't know how this can be reconciled with the theory of human origins which says that man was born from a hominid.

Do you think it is possible to restore faith in the creation of Adam and Eve in the Church?

First of all, I will repeat that the Church's position, as expressed in higher-ranking doctrinal documents, unequivocally states that Adam was formed from the dust of the earth and Eve from his side. The complete agreement of the Fathers and Doctors of the Church alone would be sufficient to recognize this truth as infallibly taught by the Church for nineteen centuries. The First Vatican Council (1868–1870) was on the verge of adopting a constitution in which this truth would have been recognized as the solemn teaching of the Church.[1] The Second Vatican Council (1962–1965) did not address this issue because it did not address any of the dogmatic teachings. Therefore, the question of the Church's ruling remains open.

Today, we are faced with an interesting situation in which the official position of the Church differs diametrically from the opinions of theologians, including many bishops. Perhaps in this context, we can understand Jesus' mysterious question: "When the Son of Man comes, will he find faith on earth?" (cf. Lk 18:8). It turns out that the Church teaches something infallibly, but its members may be confused or even tangled in errors. Nevertheless, I am convinced that this is just a transitional situation. All of the great theological renewals have been preceded by controversies and many times errors seemed to initially prevail. Even a heresy as serious as Arianism was at one point supported by a significant number of bishops. because infallibility is not granted to the Church automatically, without any human cooperation.

For centuries, theologians were aware that the Church has been entrusted with a certain deposit of faith—that is, saving truth—which the Church is to pass on intact to subsequent generations so that all the baptized may draw on the same treasures. Hence, in past centuries, there was a great awareness in the Church that she needs to maintain the doctrine intact.

Unfortunately, in the twentieth century, many theologians believed that the essential controversies regarding Catholic doctrine had already been resolved and therefore there is no need to pay much attention to them. The integrity of faith was considered a given, something no longer necessary to strive for, because it had been secured by the ancient councils. Today, all that remains to do is to work out the details, such as the form of ecumenical dialogue, the correct attitude toward nonbelievers, or Church-and-state relations. As a result, the entire attention and efforts of theologians became focused on secondary and tertiary issues. Church discussions have taken the form of sociological disputes, battles over the manner of celebrating the liturgy, ecclesiastical nomenclature, problems relating to nonbelievers—to those who believe differently and may or may not be seeking the true faith—and so on. At the same time, views alien to Christianity, such as theistic evolution, entered the Church through the back door.

Today, we Christians must realize that in the twentieth century, we overlooked one of the most important questions that can ever be asked: the question of human origins. Today, it has returned to us with full force, when, in the name of modern science, concepts derived from pagan mythologies have been proposed to replace our biblical faith. Without clarification on this issue, everything we say about humanity in other areas of Church teachings seems to hang in the air. The Church is infallibly guided by the Holy Spirit along the paths of truth. Therefore, it is to be hoped that the question of origins will also be finally clarified—for example, through the formulation of a new dogma regarding the special creation of man.

Notes

Introduction

1. Charles Darwin, On the Origin of Species by Means of Natural Selection, or the Preservation of Favoured Races in the Struggle for Life. London: John Murray, 1859.

2. The Genesis Creation Account

1. George Weigel, *Evangelical Catholicism: Deep Reform in the 21st-Century Church*. New York: Basic Books, 2013; see p. 75.
2. Jacques Monod, *Chance and Necessity,* translated by Austryn Wainhouse. New York: Vintage Books, 1972; see pp. 112–113. (Originally published as *Le hasard et la nécessité.* Paris: Seuil, 1970.)
3. Richard Dawkins, *River Out of Eden: A Darwinian View of Life.* London: Weidenfeld & Nicolson, 1995; see p. 133.

3. Of Apes and Men

1. These figures are based on the total mass (dry weight) of each element present in the human body. Calculations arrived at in other ways, such as by counting the number of atoms of each element present ("atomic percentage"), may vary considerably.
2. Ann Gauger, Douglas Axe, and Casey Luskin, eds., *Science and Human Origins.* Seattle: Discovery Institute, 2012.

3. Ibid.; see p. 41.

4. Saint Augustine, *The City of God*, Book XII, Chapter 24.

5. *Ibid.*

6. Saint Thomas Aquinas, *Summa Theologiae*, I, Q91, A2.

7. S. Joshua Swamidass, *The Genealogical Adam and Eve: The Surprising Science of Universal Ancestry.* Downers Grove, IL: IVP Academic, 2019.

8. Francisco J. Ayala, "The Myth of Eve: Molecular Biology and Human Origins," *Science,* 1995, 270: 1930–1936.

9. Ola Hössjer, Ann Gauger, and C. Reeves, "Genetic Modeling of Human History, Part 1: Comparison of Common Descent and Unique Origin Approaches," *BIO-Complexity,* 2016, 2016(3): 1–15; and Ola Hössjer, Ann Gauger, and C. Reeves, "Genetic Modeling of Human History, Part 2: A Unique Origin Algorithm," *BIO-Complexity,* 2016, 2016(4): 1–36.

4. Catholicism and Evolution

1. Michael Chaberek, *Catholicism and Evolution: A History from Darwin to Pope Francis.* Brooklyn, NY: Angelico Press, 2015.

2. The Holy Office itself continued updating the Index until 1966.

3. Pope Pelagius I, *Epistola ad Childebertum regem Francorum* [Letter to Childebert King of the Franks], in J.-P. Migne, ed., *Patrologia Latina,* 69, 405–408.

4. *Acta et decreta concilii provinciae Coloniensis in civitate Coloniensi Anno Domini MDCCCLX . . .* [Acts and decrees of the Province of Cologne, in the city of Cologne, in the Year of Our Lord 1860 . . .]. Cologne: 1862; see Title 4, Chapter 14, p. 30.

5. Are Creationists Simply Ignorant?

1. Michael Denton, *Evolution: A Theory in Crisis.* Bethlehem, MD: Adler & Adler Publishers, 1985.

2. Percival Davis and Dean H. Kenyon, *Of Pandas and People: The Central Question of Biological Origins,* edited by Charles Thaxton. Richardson, TX: Foundation for Thought and Ethics, 1989.

6. Is Evolution in Trouble?

1. For example, Robert E. Grant, Robert Jameson, William Lawrence (1st Baronet), James Cowles Prichard, and William Charles Wells.

2. Anonymous [Robert Chambers], *Vestiges of the Natural History of Creation.* London: John Churchill, 1844.

3. Lamarck's chief work was *Philosophie zoologique* [Zoological Philosophy]. Paris: Muséum d'Histoire Naturelle (Jardin des Plantes), 1809.

4. Michael Behe, *The Edge of Evolution: The Search for the Limits of Darwinism.* New York: Free Press, 2007.

5. See, for example, James A. Shapiro, *Evolution: A View from the 21st Century. Fortified.* Chippewa Falls, WI: Cognition Press, 2022.

6. Michael A. Cremo and Richard L. Thompson, *Forbidden Archeology: The Hidden History of the Human Race.* Los Angeles: Bhaktivedanta Book Publishing, 1998.

7. The communist leader Władysław Gomułka boasted: "We will never give up power once it has been taken." This phrase reflected the authoritarian stance of the communist regime in Poland, emphasizing its refusal to relinquish control once it had been established.

8. Stephen C. Meyer, *Darwin's Doubt: The Explosive Origin of Animal Life and the Case for Intelligent Design.* San Francisco: HarperOne, 2013.

9. *Darwin's Dilemma: The Mystery of the Cambrian Fossil Record* (2009), directed by Lad Allen.

10. Stephen C. Meyer, *Signature in the Cell: DNA and the Evidence for Intelligent Design*. San Francisco: HarperOne, 2009.

7. Intelligent Design—Fact vs. Fiction

1. Michael J. Behe, Darwin's Black Box: The Biochemical Challenge to Evolution. New York: Free Press, 1996.

8. Darwin and the Death Camps

1. Thomas Robert Malthus, An Essay on the Principle of Population. London: J. Johnson, 1798.
2. Charles Darwin, The Descent of Man, and Selection in Relation to Sex. London: John Murray, 1871.
3. For example, Richard Weikart, *From Darwin to Hitler: Evolutionary Ethics, Eugenics, and Racism in Germany*. New York: Palgrave/Macmillan, 2004.

9. Why Is Teaching Evolution So Controversial?

1. Oxford: Oxford University Press, 2012.
2. Jonathan Wells, Icons of Evolution: Science or Myth? Washington, DC: Regnery Publishing, 2000.

10. Was Eve Created from Adam's Rib?

1. The ratification process was cut short when Italian troops occupied the Vatican in September, 1870, following the withdrawal of French troops loyal to Pope Pius IX, in connection with France's defeat in the Franco-Prussian War.

Index

www.ingramcontent.com/pod-product-compliance
Lightning Source LLC
Chambersburg PA
CBHW060420130626
46555CB00005B/2144